COLLOIDAL POLYMER PARTICLES

COLLOIDAL POLYMER PARTICLES

EDITED BY

J. W. GOODWIN
University of Bristol
School of Chemistry, Bristol

R. BUSCALL
ICI Corporate Colloid Science Group
Runcorn, Cheshire

ACADEMIC PRESS
Harcourt Brace and Company, Publishers
London San Diego New York
Boston Sydney Tokyo Toronto

₀651800X
CHEMISTRY

ACADEMIC PRESS LIMITED
24-28 Oval Road
LONDON NW1 7DX

U.S. Edition Published by
ACADEMIC PRESS INC.
San Diego, CA 92101

This book is printed on acid free paper

A catalogue record for this book is available from the British Library

ISBN 0-12-290045-6

Printed in Great Britain

CONTENTS

PREFACE

During the past thirty years the field of polymer colloids has been a particularly fertile one with more and more interest changing from natural rubber latex to a wide variety of synthetic latices reflecting the increasing use of these materials in an increasing number of applications. During this period they have also become widely used as the model system of choice in many fundamental studies of colloidal properties whether carried out in industry or academia.

The contents of this volume reflects this wide interest with chapters ranging from the preparation and properties of conducting particles, as well as composite particles and swellable particles. The ability to thoroughly characterize polymer colloids has long been a corner-stone of their use in fundamental work and new techniques such as dielectric spectroscopy and neutron scattering are becoming increasingly important. Many other materials such as proteins, polymers, polyelectrolytes and surfactants interact strongly with the surface of polymer latex particles and new work on these aspects is well covered. Finally, latices are often used because their flow properties are suitable for their use as coatings and adhesives. The rheological contributions to this volume cover many of the recent developments.

The publication of this volume coincides with the formal retirement of Professor Ron Ottewill. Ron is well known to workers in the polymer colloid field for his wide ranging and important contributions to the field over the last thirty years and we would like to take this opportunity to mark his retirement by dedicating this volume to him.

DEDICATION

Professor R.H. Ottewill OBE FRS

During his career, there have been major advances in Colloid Science which range from new experimental techniques to theoretical understanding. During the sixties Ron helped to establish polymer latices as perhaps the best model colloidal system through his work on the stability of well characterised monodisperse latices. Subsequently he has made major contributions to the understanding of concentrated systems through his work on concentrated latices with crystalline and glassy structures. Since the seventies he has been very heavily engaged in applying neutron scattering to latex problems. In fact his contributions cover a very wide range of problems and systems so he has been a constant source of new ideas, new information and advice to most of us working in the colloid field. The diversity of his interests is extremely broad, covering areas such as surfactant behaviour, microemulsion properties, thin film and particle stability, emulsion polymerisation and colloidal crystals. Throughout his work he has always taken a problem-centred approach and brought to bear a wide range of techniques to solve each problem. With 260 papers and 8 edited books, he has provided the colloid community with a great fund of information in the general literature as well as through his efforts as a teacher.

Ron started his teaching career with a short spell at the University of London prior to going to Cambridge in 1952 to join the Department of

Colloid Science and become a Member of Fitzwilliam College. The Department was very active at that time with many major figures in the field spending time there. In 1964, after twelve years at Cambridge he brought his research group to the University of Bristol where he set up the well known one-year Masters programme which is still in place after 30 years. His research group prospered with many of his hundred plus Ph.D. students turning in theses on polymer colloid based problems. Ron became Professor of Colloid Science in 1971 and subsequently Department Head. He took the prestigious Leverhulme Chair of Physical Chemistry in 1981, a worthy successor to the line of well known encumbents since McBain in 1919. Ron has always maintained a major interest in teaching, even when he carried a high administrative load as Dean of the Science Faculty and latterly as the overall Head of School, and there are many of us who are grateful that we have had the opportunity to attend so many of his lectures.

In addition to teaching, Ron has always been a prime mover in the dissemination of information through conferences and learned societies. He is currently a Vice President and an ex-President of the Faraday Division of the Royal Society of Chemistry and has had a major involvement in the planning and running of several of the Division's Discussion meetings. Moreover he was one of the four founding members of Colloid and Interface Science Group which is now the largest subject group of that Division. During the past 25 years as committee member, secretary and finally chairman of the Group, Ron has organized a very large number of informal discussion meetings. At the beginning of the 1970s the International Polymer Colloid Group was formed with John Vanderhoff, Irvin Krieger, Bob Fitch and Ron as the founders. More recently he has helped to found the European Society, ECIS. He was Chairman of the 1987 Gordon conference on Polymer Colloids and the Director of the 1988 NATO ASI on Polymer Colloids. However Ron works hard on the dissemination of information through editorial duties on many journals. Currently he has editorial duties on Advances in Colloid and Interface Science, Colloid and Polymer Science and Langmuir. The Journal of Electroanalytical Chemistry, The Proceedings of the Royal Society (Physical Science) and the Journal of Colloid and Interface Science are previous editorial commitments.

The recognition of the many important contributions made by Ron has come from many societies and countries. In 1974 he was presented with the Surface and Colloid Medal of the Royal Society of Chemistry; in 1979 the award of the Wolfgang Ostwald medal from the Kolloid Gesellschaft in Germany; in 1981 the College de France in Paris, and in 1982 an award from Helsinki University in Finland and the Alexander Medal of the Royal

Australian Chemical Institute. He was elected to the Royal Society of London in 1982 and received the Liversidge Medal of the Royal Society of Chemistry in 1985. The most recent award has been the award of the first Colloid and Interface Science Group Medal to Ron at the Group's conference in Strasbourg in 1993. The greatest accolade, however, came in June 1990 when the Queen made Ronald Harry Ottewill an Officer of the Order of the British Empire, a fitting tribute to a most successful career.

J.W. Goodwin

POLYMER COLLOIDS IN NONAQUEOUS MEDIA

R.H. Ottewill
School of Chemistry
University of Bristol

ABSTRACT

Methods for the synthesis of spherical particles with a very narrow distribution of particle sizes have been developed over the past two decades. The monodispersity of these systems together with the spherical nature of the particles has made them ideal model systems for investigations into the many phenomena associated with colloidal systems. One particular aspect of this work has been to use scattering techniques in order to probe the structure of nonaqueous dispersions. In the systems described in this article the particles were composed of a core of poly-(methylmethacrylate) with a shell of poly-12-hydroxy stearic acid. In terms of particle-particle interaction it has been found that these particles behave as "nearly" hard spheres. Hence, depending on the volume fraction of the dispersions various phenomenological states are observed which are analogous to those seen in molecular systems, e.g. fluids, crystals and glasses. It is also demonstrated that using binary systems, crystals of the AB_2 type can be formed.

INTRODUCTION

Colloid science is a subject which is rich in the phenomena that are encountered in a wide range of every-day activities, in the academic laboratory, in industry, in biology and in the environment. A challenge to the academic scientist has been to understand in both chemical and physical terms both by experiment and by theory the fundamental basis of the behaviour of colloidal dispersions, from very dilute to very concentrated and even into the dried state. In order to achieve this goal experiments need to be precise and an interpretation carried out in detail so that a comparison can be made with theoretical models. In the development of theoretical models, mathematical manipulation is aided substantially by spherical symmetry of the particles and by the fact that in an ensemble of particles it can be assumed that all the spheres have the same diameter, or a very narrow distribution of particle sizes. These factors place a large premium on the chemist being able to synthesize in the laboratory spherical particles with a narrow distribution of diameters and a surface of known properties. In fact, over the last thirty years or so, considerable effort has been devoted to this subject, in particular, to the preparation of particles in which the basic material is an organic polymer; these are now frequently termed polymer colloids. In aqueous environments many different types of polymeric particles have been produced, those prepared most frequently in academe probably being polystyrene. However, the prepatation of polymer colloids in nonaqueous media has not been studied as extensively as that with water as the dispersion medium.

In a number of ways the nonaqueous environment has an advantage over the aqueous environment. These include the extensive variety of liquids available covering a wide range of properties from apolar to polar and, for example, in scattering studies the possibility to vary the refractive index of the medium, and even to match the refractive index of the dispersed particles by the use of liquid mixtures. Also, the potential energy of interaction for sterically stabilised particles in the nonaqueous environment is often very close to being that of a hard sphere. Consequently a wide range of theory developed to explain the physics of fluids becomes potentially available for transposition.

This lecture is a rather personal one and in it, I will try and review some of the work carried out in Bristol in the last two decades or so on polymer colloids in non-aqueous dispersions.

Synthesis of Polymer Colloids in Nonaqueous Media

A critical pathway to examining the properties of colloidal dispersions at a fundamental level is the synthesis of colloidal stable, monodisperse particles of well-defined shape, at high volume fractions in a consistent and reproducible fashion. A corollary to this is to have a detailed knowledge of the particle morphology and, in particular, a knowledge of the surface and the manner in which it controls the stability of the dispersion. It is also an advantage to be able to control the mean particle diameter of the particles in order to make available a wide range of sizes.

Historically, one of the first significant attempts to prepare sterically stabilised polymeric particles was the work of Osmond and Thompson (1). An account of this early work has been given by Barrett (2). These authors used methyl methacrylate as the monomer in a hydrocarbon solution of degraded rubber with benzoyl peroxide as the initiator; this enabled an amphipatic copolymer to be formed in situ and then grafted to the particle. Early work at Bristol (3) utilised the formation of seed particles followed by a monomer feed process to grow the particles in the presence of a polymeric stabiliser. However, in later work (4) this was superceded by a "single-shot" method which was found to be capable of producing very monodisperse systems and also of providing a wide range of particle diameters. A key ingredient in this work was the use of a stabiliser molecule in the form of a "comb", that is with a backbone composed of a copolymer of poly-(methyl methacrylate) (PMMA) and poly(glycidyl meth-acrylate) and "teeth" of poly(12-hydroxy stearic acid) (PHS), In most preparations the "comb" backbone was subsequently covalently linked to the particle core of poly(methyl methacrylate); this prevented the possibility of desorption of the polymeric stabiliser during interaction with other particles. An illustration of this type of particle is shown schematically in Figure 1; it corresponds to a core particle of radius R with a shell of thickness, δ. The latter is essentially the extended length of the PHS molecule, ca. 10 nm, and even though the core particle has been varied over wide limits, there has been

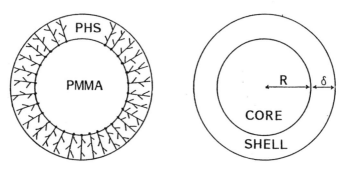

Figure 1:   Schematic illustration of a PMMA core particle of radius R stabilised by a PHS shell of thickness δ

substantial agreement between various authors that the thickness of the PHS layer is essentially constant at ca. 9 ± 1 nm (3-8).

The mechanism of synthesis of PMMA-PHS particles in a hydrocarbon (hexane-dodecane) - monomer environment is illustrated schematically in Figure 2. According to the procedure previously described (4) the initial monomer concentration in the liquid mixture is adjusted to be between 35 and 50% depending on the final particle size required.   Initiation was achieved using azobisisobutyronitrile (ABIN) at a temperature of 80°C. As monomer is converted to polymer following initiation the polarity of the medium decreases to a point where the polymer chains essentially become insoluble;   hence precipitation occurs with the consequent formation of small poly(methylmethacrylate) particles.   Evidence suggests that initially these are not colloidally stable and coagulation occurs to form larger paticles.   However, as the polarity of the medium changes conditions also become favourable for the adsorp-

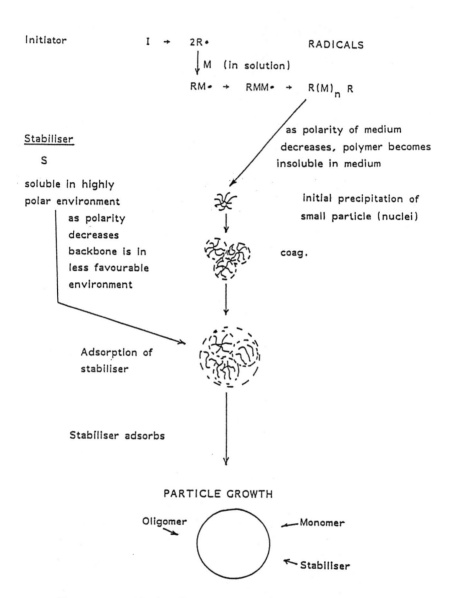

Initiator   I → 2R•   RADICALS

                              │ M (in solution)

                              RM• → RMM• → R(M)ₙ R

                                              as polarity of medium
Stabiliser   decreases, polymer becomes
                                              insoluble in medium
     S

soluble in highly   initial precipitation of
polar environment   small particle (nuclei)

        as polarity
        decreases
        backbone is in   coag.
        less favourable
        environment

    Adsorption of
    stabiliser

    Stabiliser adsorbs

                        PARTICLE GROWTH

                Oligomer   ⟨ ⟩   —Monomer

                                    ←Stabiliser

**Figure 2: Mechanism of formation of PMMA-PHS
particles (schematic)**

tion of stabiliser molecules on to the particles; hence at
this stage a stable colloidal dispersion is formed. This
stage appears to control the number concentration and for
the rest of the reaction, these particles continue to grow
until the monomer has been consumed (2,3,4,9). The particles formed by this method generally have coefficients of
variation on the mean diameter of less than 10%.

The PMMA-PHS particles form stable dispersions in
aliphatic hydrocarbon media such as dodecane and in poly-
cyclic saturated hydrocarbons such as decalin. The
colloidal stabilisation of the particles is provided by the PHS
chains in the outer shell of the particles. This suggests
that the mechanism is steric stabilisation as indicated schematically in Figure 3. As shown later in this review and in
previous investigations (3) the evidence suggests that the
particles are not quite hard spheres (see Figure 3) since
the shell is slightly compressible. Nevertheless, within
the limits of current experimentation it appears reasonable
to treat them as having a "nearly hard-sphere" behaviour.

THE STRUCTURE OF CONCENTRATED COLLOIDAL DISPERSIONS

The preparation of monodisperse polymer colloid dispersions has led to many applications. One of the major
ones in academic studies has been to use them as model
systems to probe the behaviour of concentrated colloidal
dispersions. An initial task in this respect is to understand the correlations which occur between particles as
the system becomes more and more concentrated. In

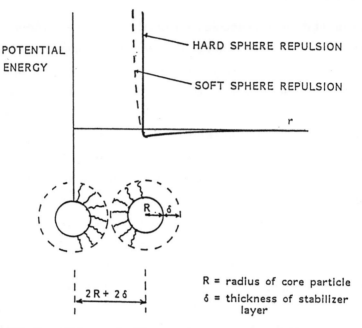

Figure 3: Schematic illustration - steric interaction between PMMA-PHS particles to compare hard sphere interactions with interactions between slightly soft spheres

simple terms this means to determine the structural arrangement of the particles in a medium and to understand how this depends on variables such as, the pair interaction potential, the relative permittivity of the medium, the concentration of salt present etc. In physical terms this means determining the pair correlation function $g(r)$ defined by (12), the following equation,

$$g(r) = \frac{N(r)}{N_p} = 1 + \frac{1}{2\pi^2 r N_p} \int_0^\infty [S(Q) - 1]$$

$$Q \sin Qr \, dQ \qquad \ldots \quad (1)$$

Essentially g(r) is the probability of finding another
particle at a distance r from a reference particle, where N_p
is the average number of particles in the system per unit
volume and N(r) is the number density as a function of the
distance parameter r. The other terms included in equation (1) include the scattering vector Q, which has dimensions of reciprocal length and the structure factor S(Q).
The latter quantity can be obtained by Fourier transformation of equation (1), giving

$$S(Q) = 1 + \frac{4\pi N_p}{Q} \int_0^\infty r\,[g(r) - 1]\,\sin Qr\,dr \quad \ldots \text{(2)}$$

S(Q) is a quantity of major importance since it can be
obtained from scattering experiments on dispersions (13,14).

SCATTERING EXPERIMENTS

The three forms of radiation most widely used for
scattering experiments are light, X-rays and neutrons. In
terms of wavelength light covers the range ca. 400 to 650
nm, X-rays ca. 0.04 to 0.5 nm and cold neutrons the
range from ca. 0.5 to 2.0 nm. Each of these has advantages and disadvantages but the present review will deal
primarily with neutron beams. The fact that distances
from ca. 1 nm to 1 μm can be probed makes this an
attractive method for colloidal dispersions. In a neutron
scattering experiment the intensity of scattering is measured over a small range of angles in the forward direction,
usually up to about 15°, and for a single spherical particle of volume V_p the intensity of scattering is expressed

as,

$$I(Q) = A \ V_p^2 \ (\rho_p - \rho_m)^2 \ P(Q) \qquad \qquad \ldots \quad (3)$$

where A is an instrumental constant, ρ_p = the coherent scattering length of the particle, ρ_m = the coherent scatterint length of the medium and P(Q) is the particle shape factor, which for a spherical particle of radius R is given by (15),

$$P(Q) = \left(\frac{3(\sin QR - QR \cos QR)}{(QR)^3} \right)^2 \qquad \ldots \quad (4)$$

The scattering vector $\underline{Q}$ for elastic scattering has a magnitude given by

$$Q = 4\pi \sin (\theta/2)/\lambda \qquad \qquad \ldots \quad (5)$$

where λ = the wavelength of the incident radiation and θ = the scattering angle. For core-shell particles the particle shape factor has a slightly more complex form but is readily accessible (16).

In the case of concentrated dispersions equation (3) needs to be modified to take into account the higher number density and interparticle correlations. This leads to,

$$I(Q) = A \ N_p \ V_p^2 \ (\rho_\rho - \rho_m)^2 \ P(Q) \ S(Q) \qquad \ldots \quad (6)$$

Since N_p is expressed as the number of particles per unit volume, $N_p V_p$ can be replaced by the volume fraction, namely, $\phi = N_p V_p$.

In practice the volume fraction can be determined, ρ_p and ρ_m are known and A can be determined. Hence equation (3) can be put into the form,

$$P(Q) = \frac{I(Q)}{A\,\phi\,V_p\,(\rho_p - \rho_m)^2}$$

Since $V_p = 4\pi R^3/3$, the radius R can be determined by electron microscopy or preferably by a scattering experiment. For a polymer colloid dispersion, even one given the accolade "monodisperse" there will be a distribution of particle sizes and hence a size distribution function has to be introduced (17). An example of the results obtained on a dilute dispersion of PMMA-PHS particles (R = 15 nm) in dodecane is given in Figure 4. As can be seen there are a number of points obtained over a wide range of Q and hence it is possible to fit such a curve and to obtain, using a log-normal distribution function (18), a particle size distribution and the form of P(Q) against Q (17).

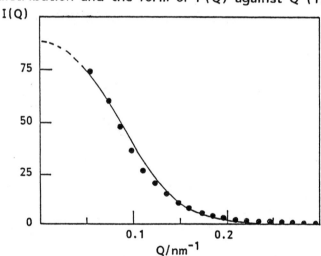

Figure 4: I(Q) in arbitrary units against Q for PMMA-PHS particles (R_c = 15 nm) in dodecane at a volume fraction = 0.023; ———— , theory

Results for the same PMMA-PHS dispersion at high volume fractions are illustrated in Figure 5 as curves of intensity vs Q. Thence using equation (6) over a range of Q the results can be converted directly into the form S(Q) against Q. Results in this form are illustrated in Figure 6. These results show clearly the difference between dilute dispersions and concentrated dispersions. A well-defined, but broad peak, is present which moves to higher Q values as the volume fraction is increased.

Fourier transformation of the S(Q) versus Q data leads directly to the pair correlation function g(r) as a function of r and examples are shown in Figure 7. These curves have an oscillatory form with a clearly defined peak

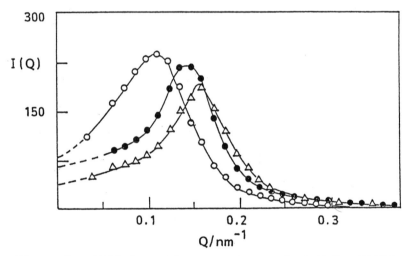

Figure 5: I(Q) in arbitrary units against Q for PMMA-PHS particles (R_0 = 15 nm) in dodecane at high volume fractions: O , 0.23; ● , 0.36; △ , 0.42.

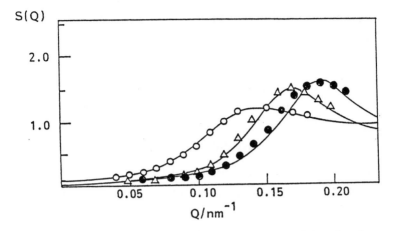

Figure 6: S(Q) against Q, PMMA-PHS particles in dodecane, R_c = 15 nm at various volume fractions: ○ ,
0.23; △ , 0.36; ● , 0.42. ——— , fit to hard-sphere
model (11)

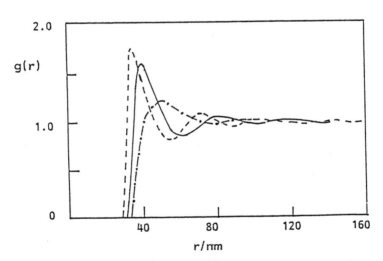

Figure 7: g(r) against r, PMMA-PHS particles in dodecane,
R_c = 15 nm at various volume fractions; — • — • , 0.23;
——— , 0.36; — — — —, 0.42

at a distance of the order of 40 nm which, as the volume
fraction increases, becomes larger in magnitude and
narrower in width; the first maximum also moves to a smaller r value.   At the highest volume fraction secondary, tertiary and quarternary peaks are visible.   These results
suggest that treated as a one-component system, in which
the particles can be considered, by analogy to be behaving
as molecules, then the particle distribution resembles the
fluid state.   That is, there is short-range order and long-
range disorder.   It can be anticipated therefore that the
particles should be in diffusional motion even at high-
volume fractions.   Indeed direct measurements of the long-
time self-diffusion of PMMA-PHS particles indicates that it
occurs up to a volume fraction of ca. 0.5 when it ceases
(Figure 8).   At this volume fraction for systems with a
coefficient of variation on the mean particle size of less
than ca. 10% the particles start to form crystalline arrays
(19-22).   It is of considerable interest that Hoover and
Ree (23) predicted that for a hard sphere system freezing
should occur at a volume fraction of 0.495 and a melting
transition at a volume fraction of 0.545.

The g(r) against r curves illustrated in Figure 7
suggest that the PMMA-PHS particles are not ideal hard-
spheres in the statistical mechanical sense.   There is, for
example, a small positive gradient, in the curves at the
smallest r values;   moreover, the fact that the onset of
repulsion moves to smaller r values as the volume fraction
increases would tend to suggest that the stabilising shell
may become slightly more compressed as the volume fraction increases.   Essentially the repulsion becomes harder

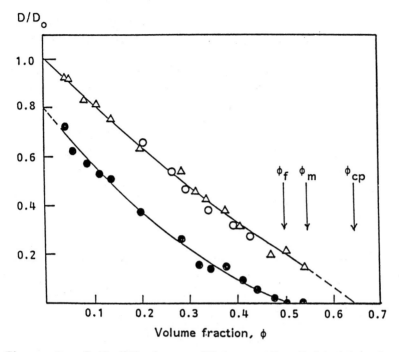

Figure 8: Self-diffusion coefficients, D, divided by the translational diffusion coefficient of a single particle, D_o for PMMA-PHS particles in dodecane: ● , long-time self-diffusion coefficient, particle diameter = 166 nm; open symbols, short-time self-diffusion coefficients, △ , particle diameter = 166 nm, O , particle diameter = 1.18 μm

as the onset of crystallisation is approached and some solvent molecules are squeezed out of the outer layer, potentially gaining in entropy as they return to the dispersion medium; this point means that care has to be exercised in neglecting the dispersion medium and treating it as a structureless continuum.

A further point of interest can be made from the translational diffusion measurements reported in Figure 8, namely that the short-time diffusional motion does not cease

at $\phi \sim 0.5$ but continues beyond that point (24), indeed
it continues through the region of crystallisation. A tenta-
tive extrapolation of the results suggests that the short time
self-diffusion, which must essentially be the limited motion
in a "cage" of other particles, ceases at ca. $\phi = 0.64$, the
volume fraction of random close-packing of spheres. It is
approximately at this volume fraction that a colloidal glassy
state is formed, namely, a high volume fraction of particles
packed into an amorphous array. The form of $g(r)$ against
r for such a system closely resembles those shown in
Figure 7. Although the glassy state can often be recognised visually by the failure to observe the Bragg diffraction of visible light at high volume fractions, it is difficult
to recognise from time-average measurements, e.g. $S(Q)$
against Q. It is, however, readily apparent from dynamic
measurements, e.g. dynamic light scattering (25).

In Figure 9 a schematic diagram is presented in an
attempt to summarise the various states which can be examined using PMMA-PHS particles as "nearly" hard spheres.
In dilute dispersions, say $\phi < 0.01$, the particles are in
free Brownian motion and only interact on the occasional
collision. It is tempting to draw the analogy to the gaseous
state but it must be remembered that motion of molecules
in a gas is ballistic but particle motion in a medium is
Brownian. However, the form of $g(r)$ against r will be
very similar in the two cases. As the volume fraction increases the interactive collisions will increase in number
until on a time-average basis the system will have a "fluid-
like" structure of the type revealed by the scattering
measurements and shown as $S(Q)$ in Figure 6.

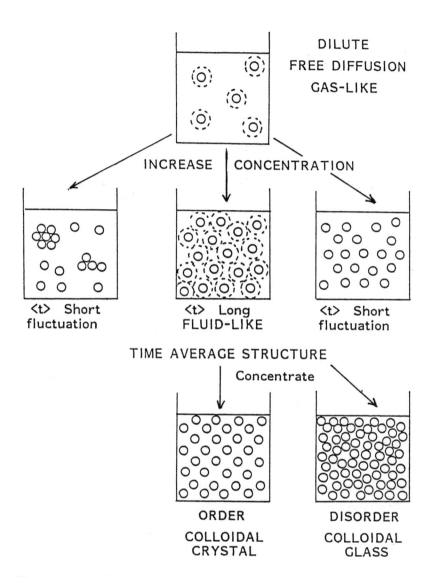

Figure 9: Schematic illustration of the various types of structure in dispersions of PMMA-PHS monodisperse particles

A probe of the system on a short time scale, since the particles are still in motion, will reveal fluctuations in number
density.   As a consequence, as in a liquid, there will be
regions with a higher number density and regions with a
lower number density.   It is also possible on a short time
scale that the fluctuations could lead to a situation close
to a regular array for short periods (Figure 9).

As the volume fraction increases further the smaller
regions of more densely packed particle clusters form
nuclei which generate crystals of the particles.   It can be
anticipated that at the onset of crystallization, a multi-
crystalline assembly will form;   subsequent rearrangement
may then with time lead to rearrangement to form larger
crystals (19).   However, as short-time diffusional motion
becomes smaller, this will become increasingly difficult
as the glassy state is approached.   Moreover, as shown by
van Megen and Pusey (19) crystallization can also occur
slowly from the glassy state.

The formation of colloidal crystals has now been
widely studied and many other systems show similar phenomena in both aqueous and nonaqueous media including
polystyrene latices (26,27), fluorinated polymer colloids
(28,29) and silica particles coated with octadecanol and
with silane coupling agents (30).

MIXTURES OF COLLOIDAL PARTICLES

It follows directly from equation (3) and (6) that if
the coherent scattering length of the medium, ρ_m, is made

the same as the coherent scattering length of the particle, ρ_p, then in principle the scattering intensity should be zero. In practice under these conditions zero intensity is not quite achieved since there is usually some small heterogeneities in the particle and also some incoherent scattering. Nevertheless under these conditions a low intensity is achieved and the particles can essentially be contrast matched (31).

This physical situation leads to the possibility of examining binary mixtures of particles, e.g., A particles and B particles, such that ρ_A is different from ρ_B but either one or each of the particles in turn can be examined in a contrast matched situation. In recent work (8) we have taken A particles, prepared from h_8-methylmethacrylate, giving $\rho_A = 1.07 \times 10^{10}$ cm^{-2} and an average diameter of 331 nm and B particles prepared from d_8-methylmethacrylate giving $\rho_B = 7.02 \times 10^{10}$ cm^{-2} with an average diameter of 93 nm. This gives a ratio for the diameter of the B particles to that of the A particles of 0.31.

It then follows from equation (6) for the situation, $\rho_m = \rho_A$, that

$$I(Q)_B = A \, N_B \, V_B^2 \, (\rho_B - \rho_m)^2 \, P_B(Q) \, S_{BB}(Q)$$

and for $\rho_m = \rho_B$

$$I(Q)_A = A \, N_A \, V_A^2 \, (\rho_A - \rho_m)^2 \, P_A(Q) \, S_{AA}(Q)$$

For an intermediate situation $\rho_m \neq \rho_A \neq \rho_B$ then

$$I(Q) \propto (\rho_A - \rho_m)^2 \, P_A(Q) \, S_{AA}(Q) + (\rho_B - \rho_m)^2 \, P_B(Q) \, S_{BB}(Q)$$
$$+ 2 \, (\rho_A - \rho_m)(\rho_B - \rho_m)[P_A(Q) \, P_B(Q)]^{\frac{1}{2}} \, S_{AB}(Q)$$

Thus from a series of experiments on a binary system it

becomes possible to extract the partial structure factors, $S_{AA}(Q)$, $S_{BB}(Q)$ and $S_{AB}(Q)$.

Experiments were carried out using small angle neutron scattering of a binary system with ϕ_A = 0.540, ϕ_B = 0.067 giving ϕ_{total} = 0.61 and N_B/N_A = 4.2 ± 0.4. It was thus anticipated that the A particles would crystallise with the B particles taking up interstitial positions in the lattice. In this mixture significant crystallisation was observed after a few hours and after 24h the sample volume was filled with small crystallites. The curves of scattered intensity using mixed liquids of different scattering lengths namely mixtures of d_{18}-octane and h-cis-decalin, are shown in Figure 10 (8). At the smallest value of ρ_m, -0.03 x

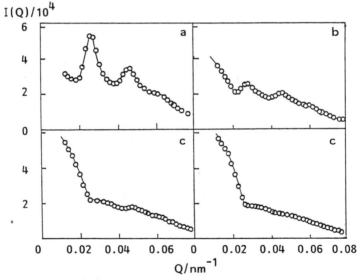

Figure 10: I(Q) against Q for a bimodal mixture of PMMA-PHS particles in d_{18}-octane-h_{18}-cis-decalin mixture to give ρ_m values; a) -0.03 x 10^{10} cm^{-2} b) 0.63 x 10^{10} cm^{-2} c) 1.02 x 10^{10} cm^{-2} d) 1.24 x 10^{10} cm^{-2}. ———, curves calculated using partial structure factors given in Figure 11

10^{10} cm^{-2} the scattering is dominated by the large particles in the lattice and 2 diffraction peaks are evident. As
ρ_m is increased then the scattering from the large A particles decreases and at ρ_m = 1.24 x 10^{10} cm^{-2} the scattered
intensity is predominantly from the B particles. It should
be noted that contrast match did not occur exactly at
ρ_A = 1.07 x 10^{10} cm^{-2} since some preferential occlusion of
small amounts of d$_{18}$-octane into the PHS layer on the
particles occurred.

In all five series of experiments were carried out at
various values of ρ_m giving five equations from which to
extract the three quantities, $S_{AA}(Q)$, $S_{BB}(Q)$ and $S_{AB}(Q)$.
The three partial structure factors are illustrated in Figure
11. Figure 11a shows the structure factor $S_{AA}(Q)$ for the
A particles, and this shows clearly two peaks, one at Q =
0.026 nm^{-1} and the other at Q = 0.046 nm^{-1}, indicating
strong spatial correlations between the particles as expected in a crystalline structure. The peak for the interplane
(001) reflection was expected to occur at 0.023 nm^{-1} and
the (111) and (002) reflections at 0.044 m^{-1} and 0.046 nm^{-1};

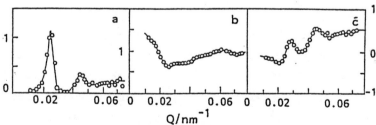

Figure 11: Partial structure factors for a bimodal mixture
of PMMA-PHS particles a) $S_{AA}(Q)$ b) $S_{BB}(Q)$
c) $S_{AB}(Q)$

the resolution of the neutron scattering instrument (D11 at the Institut Laue Langevin) would not have resolved the difference between the latter two peak positions. Clearly, the A particles appear to be in a crystalline array.

The results for $S_{BB}(Q)$, however, shown in Figure 11b are quite different and no strong peaks are apparent. There is a rather diffuse peak at $Q \approx 0.064$ nm^{-1} which is close to $2\pi/d_B$ (0.067 nm^{-1}), as expected for the one component BB structure factor. The main effect occurs at the low Q values with a minimum at $Q = 0.025$ nm^{-1} and a steep rise occurs as Q decreases which seems to peak just below $Q = 0.02$ mn^{-1}. This seems to correspond to an inhomogeneity on the scale of the large sphere $(2\pi/311 \approx 0.02$ nm$^{-1})$. These results seem to be in reasonable agreement with the predictions of the Biben and Hansen (32) equation for small spheres in the presence of larger ones.

In the case of hard sphere interactions, Vrij (33,34) has given expressions for scattering by multicomponent mixtures within the limits of the Percus-Yevich approximation. This approach when applied to the binary mixture used, and smeared by the instrumental resolution, seems to give a reasonable interpretation of $S_{AB}(Q)$. As can be seen $S_{AB}(Q)$ shows the two peaks from the crystal of the A particles and the rising peak at low Q indicates the presence of the B particles. These results appear to indicate that the B particles are present as a fluid phase coexisting with the crystalline A phase rather than going into well-organised interstitial positions.

Evidence for superlattice formation, however, was obtained with PMMA-PHS dispersions using particles of diameter, d_A = 642 nm and d_B = 372 nm, giving d_B/d_A = 0.58. An AB_2 structure was observed in dispersions, with number ratio's N_B/N_A of between 4 and 6 and total volume fractions $(\phi_A + \phi_B)$ of 0.525 and 0.557. Some weeks after preparation the homogeneous dispersion separated into a lower solid phase containing crystallites and an upper fluid phase. The dried solid phase was examined using scanning electron microscopy and an electron micrograph of this is shown in Figure 12. This shows clearly the AB_2 structure consisting of alternating hexagonal layers of large and small spheres. A super lattice structure of AB_{13} was also found in dispersions with an excess of small spheres at number ratio's N_B/N_A of 9, 14, 20 and 30 with ϕ_A and ϕ_B in the range 0.512 to 0.533. The crystallisation was much more rapid than with AB_2, crystals appearing within a few days. The

Figure 12: Scanning electron micrograph showing the formation of an AB_2 crystal lattice composed of PMMA-PHS particles

structure was confirmed by optical diffraction measurements but was found more difficult to observe by electron micro-scopy (35).

APPLIED FIELD

In addition to the experiments described in the preced-ing sections it has been possible using small angle neutron scattering to examine changes in the structure of PMMA-PHS dispersions in a nonaqueous environment brought about by the application of external fields to the system. These will only be referred to briefly. Using a Couette viscometer constructed from quartz, dispersions were examined at a volume fraction of ca. 0.4, over a range of shear rates from 300 s^{-1} to 10000 s^{-1}. The results obtained suggested that at high shear rates the particles moved further apart in the direction of flow but moved closer together in a direction perpendicular to the flow direction possibly as a consequence of cluster formation in the vorticity direction (36,37).

On the addition of calcium octanoate to a dodecane dispersion of PMMA-PHS particles a change of structure was produced in the system since a weak electrostatic charge was conferred to the particles. Application of an electric field to these systems showed the formation of strings of particles between the electrodes (38). The form of the strings seemed to bear a resemblance to the clusters formed under a shear field.

CONCLUSION

The synthesis of PMMA-PHS particles in a monodisperse and very concentrated form has led to the discovery of a rich array of colloidal phenomena of a fundamental nature. Much, however, still remains to be discovered with these systems and as pointed out by Perrin (39) many years ago, there are many analogies between the behaviour of particles and the behaviour of molecules. These factors emphasize the importance of Colloid Science as a continuing area of fundamental research.

ACKNOWLEDGEMENTS

During my twenty-eight years at Bristol it has been my good fortune to have many able and enthusiastic collaborators in the work described. In particular, I should like to express my thanks to Paul Bartlett, Jim Goodwin, Ian Livsey, Ivana Marković, Sara Papworth, Peter Pusey, Adrian Rennie, Sylvia Underwood, Julian Waters and Neal Williams.

REFERENCES

1. Osmond, D.W.J. and Thompson, H.H. (1962). British Patent 893,429.

2. Barrett, K.E.J. (1975). *Dispersion Polymerisation in Organic Media*, Wiley, London.

3. Cairns, R.J.R., Ottewill, R.H., Osmond, D.W.J. and Wagstaff, I. (1976). J. Colloid Interface Sci., 54 45-51.

4. Antl, L., Goodwin, J.W., Hill, R.D., Ottewill, R.H., Owens, S.M. and Papworth, S. (1986). Colloids and Surfaces, 17, 67-78.

5. Cebula, D.J., Goodwin, J.W., Ottewill, R.H. Jenkin, G. and Tabony, J. (1983). Colloid and Polymer Science 261, 555-564.

6. van Megen, W., Ottewill, R.H., Owens, S.M. and Pusey, P.N. (1985). J. Chem. Phys., 82, 508-515.

7. Livsey, I. and Ottewill, R.H. (1991) Adv. Colloid and Interface Sci., 36, 173-184.

8. Bartlett, P. and Ottewill, R.H. (1992). J. Chem. Phys., 96, 3306-3318.

9. Croucher, M.D. and Winnik, M.A. (1990). In *An Introduction to Polymer Colloids* (Ed. F. Candau and R.H. Ottewill), 35-72, Kluwer, Dordrecht.

10. Cairns, R.J.R., van Megen, W. and Ottewill, R.H. (1981). J. Colloid Interface Sci., 79, 511-517.

11. Marković, I., Ottewill, R.H., Underwood, S.M. and Tadros, Th.F. (1986). Langmuir, 2, 625-630.

12. McQuarrie, D.A. (1976). Statistical Mechanics, 254-289, Harper and Row, New York.

13. Brown, J.C., Pusey, P.N., Goodwin, J.W. and Ottewill, R.H. (1975). J. Phys. A., 8, 664-682.

14. Cebula, D.J., Goodwin, J.W., Jeffrey, G.C., Ottewill, R.H., Parentich, A. and Richardson, R.A., (1983). Faraday Discuss. Chem. Soc., 76, 37-52.

15. Guinier, A. and Fournet, G. (1955). *Small Angle Scattering of X-rays*. Chapman and Hall, London.

16. Marković, I., Ottewill, R.H., Cebula, D.J. Field, I. and Marsh, J.F. (1984). Colloid and Polymer Science, 262, 648-656.

17. Marković, I. and Ottewill, R.H. (1986). Colloid and Polymer Science, 264, 65-76.

18. Espenscheid, W.F., Kerker, M. and Matijević, E. (1964). J. Physical Chem., 68, 3093-3097.

19. Pusey, P.N. and van Megen, W. (1986). Nature, 320, 340-342.

20. Barrat, J.L. and Hansen, J.P. (1986). J. Physique, 47, 1547

21. Pusey, P.N. (1987). J. Physique, 48, 709-712.

22. Ottewill, R.H. (1989). Langmuir, 5, 4-11.

23. Hoover, W.G. and Ree, F.H. (1968). J. Chem. Phys., 49, 3609-3617.

24. Ottewill, R.H. and Williams, N.St.J. (1987). Nature, 325, 232-234.

25. van Megen, W., Underwood, S.M., Ottewill, R.H., Williams, N.St.J., Pusey, P.N. (1987). Faraday Discuss. Chem. Soc., 83, 47-57.

26. Ashdown, S., Marković, I., Ottewill, R.H., Lindner, P., Oberthür, R.C. and Rennie, A.R. (1990). Langmuir, 6, 303-307.

27. Hachisu, S., Kobayashi, Y. and Kose, A. (1973). J. Colloid Interface Sci., 42, 342.

28. Ashown, S. (1990). Ph.D. thesis, University of Bristol.

29. Ottewill, R.H. (1990). Faraday Discuss. Chem. Soc., 90, 1-15.

30. Smits, C., Briels, W.J., Dhont, J.K.G. and Lekkerkerker, H. (1989), Prog. Colloid and Polymer Science, 79, 287-292.

31. Ottewill, R.H. (1981). In *Colloidal Dispersions* (Ed. J.W. Goodwin), 143-163. Royal Society of Chemistry, London.

32. Biben, T. and Hansen, J.P. (1990). Europhys. Lett.,
    12, 347–

33. Vrij, A. (1978). J. Chem. Phys., 69, 1742–

34. Vrij, A., (1979). J. Chem. Phys., 71, 3267–

35. Bartlett, P., Ottewill, R.H., Pusey, P.N. (1992).
    Phys. Rev. Lett., 68, 3801-3804.

36. Lindner, P., Marković, I., Oberthür, R.C.,
    Ottewill, R.H. and Rennie, A.R. (1988). Prog. Coll-
    oid and Polymer Sci., 76, 47-50.

37. Ottewill, R.H. and Rennie, A.R. (1990). Int. J. Multi-
    phase Flow, 16, 681-690.

38. Ottewill, R.H., Rennie, A.R. and Schofield, A. (1990).
    Prog. Colloid and Polymer Sci., 81, 1-5.

39. Perrin, M.J. (1910). *Brownian Movement and Molecular
    Reality*, Taylor and Francis, London.

CRYSTALS MADE OF CLOSE PACKED POLYMERIC SPHERES A SMALL ANGLE NEUTRON SCATTERING STUDY[*]

J. Rieger, O. Dippel, E. Hädicke, G. Ley
BASF AG, Polymer Research Division, 6700 Ludwigshafen, Germany
P. Lindner
Institut Laue Langevin, 38042 Grenoble, France

Abstract

Small angle neutron scattering experiments were performed on latex films prepared from a dispersion consisting of monodisperse spherical polymeric particles. Analysis of the scattering curves leads to the following results:

- The latex films consist of ordered domains (crystallites) where the polymeric spheres are close-packed in a face centered cubic structure.

- The intersticies between the close-packed spheres are filled with non-polymeric material such as salt, emulgator, etc. and possibly micro-voids. When the dried films are exposed to water (D_2O) part of this water diffuses into the intersticies.

- When annealing films at elevated temperatures additional non-polymeric material (surfactants) diffuses into the intersticies.

[*] Dedicated to Prof. Ron Ottewill on occasion of his formal retirement

1. Introduction

Latices are colloidal dispersions consisting of discrete - mostly spherical - polymer particles in a fluid phase. Latex films are prepared from the dispersion by evaporation of the liquid. During the process of film formation three subsequent processes can be discerned, as is schematically illustrated in Fig. 1.:

1. Increasing concentration of the polymeric material during evaporation of the fluid results in a dense packing of the particles (Fig. 1a and 1b).

2. Deformation of the particles to polyhedra and beginning coalescence because of surface- and capillar-forces (Fig. 1c).

3. Further coalescence caused by the interdiffusion of polymeric chains emanating from contiguous particles (Fig. 1d).

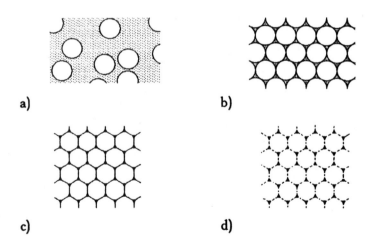

a)   b)

c)   d)

Fig. 1
Stages during the process of film formation.

The three processes sketched in Fig. 1 have been examined experimentally as well as theoretically. The reader is referred to Refs. 1 - 4 (point 1), Refs. 4 - 7 (point 2), and Refs. 8 - 10 (point 3). But, still there are many open questions, e.g.:

1. Which structure has evolved during film formation?
2. How is this structure affected by different film formation temperatures?
3. How far has the coalescence of the particles already proceeded in a film with a given thermal history?
4. To which degree is the process of coalescence homogeneous on spatial length scales?

In order to obtain more knowledge about the structure of latex films we performed small angle neutron scattering (SANS) experiments on latex films. The films prepared differed with respect to their thermal history and with respect to the amount of water which was brought into the films after drying. Our investigations differ from published experiments [4] insofar as we brought the water (D_2O) into the respective samples <u>after</u> completed formation of the film. Using this approach we obtain information with respect to the above questions, which are interesting for fundamental reasons as well as for the application of industrially produced latices in paints for buildings and cars, in surface refining of paper, etc.

In the next section the latex systems used, the preparation of the samples, as well as experimental details will be described. Section 3 provides a model for the interpretation of the measured scattering curves. Further information extracted from the data is discussed in Section 4.

2. Samples and experimental setup

The dispersion from which the films were prepared is a styrene/ n-butylacrylate copolymer latex with about 50% by weight of each monomer and a polymer content of about 50% by weight. The system is stabilized against coagulation by the existence of negative surface charges on the latex particles originating from sulfate groups. The dispersion contains additional emulsifier and electrolyte. The sample was characterized in the usual manner by light scattering and electron microscopy. The particle size distribution was quite narrow; ultracentrifugation resulted in the trias d10/d50/d90 of 124/129/134 nm, respectively. The particles can be considered as being spherical. The latex was mixed with an equal amount of destilled water and samples of the diluted latex were spread on flat silicone dishes in an amount to yield after evaporation of water latex films of about 500 μm thickness. Films were prepared by the evaporation of water at room temperature (climatized room 20^0C, 60% rel. humidity) and 70^0C, respectively. The glass temperature of the films is at about room temperature. In order to prepare the samples for SANS measurements the latex films were cut into pieces fitting into the sample holders used at the SANS instruments. Every two such film pieces of known mass were packed face to face into a little bag made from polyethylen foil together with the amount of D_2O wanted and the bag was sealed off. D_2O instead of H_2O was used because of the enhanced scattering intensity. SANS samples prepared as described above were equilibrated at least 10 days in the climatized room before measurement. The sealed off samples afterwards were stored and handled at room temperature. During the first days of equilibration and storage time and after having performed the SANS measurements the mass of the samples was controlled. It was found that the samples after the SANS measurements had lost small amounts of water. This effect is presumably due to uncomplete sealing and/or permeation of water through the poly-ethylene foil.

The SANS experiments were performed at the D 11 instrument of the high flux reactor at the Institut Laue Langevin (Grenoble). The sample to detector distance was chosen to be 20m. The wavelength of the neutrons was 1nm with a wavelength distribution of $\Delta\lambda/\lambda=$ 0.09. The data was sampled and manipulated in the standard way in order to obtain the scattering curves $I(q)$ [11]. Fig. 2 shows a typical example of intensities sampled by the two-dimensional detector of the D 11 instrument. Diffraction of the powder-like (Debye-Scherrer) type is observed. The symmetry of the data is exploited for averaging and the determination of the radial distribution function $I(q)$ [11]. $I(q)$ denotes the intensity of neutrons scattered at an angle Θ with respect to the beam sent onto the sample. q is defined by $q=4\pi\sin(\Theta/2)/\lambda$.

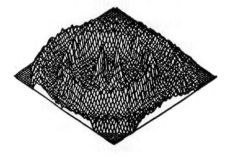

Fig. 2

Example of the scattering intensity as sampled by the two-dimensional detector (raw data). The examined film was prepared at room temperature and contained $\sim 2\%$ D_2O. The length of the square corresponds to $\sim 0.18\,nm^{-1}$.

No attempt was made to determine absolute values for the intensities since the films contained many air bubbles with a size of about 500 µm which prevented the determination of the effective density of the samples. Experiments were performed with different sample thicknesses in order to check for the possible occurrence of multiple scattering. No such effects were observed.

3. Interpretation of the scattering data I(q)

The model. A typical set of data is shown in Fig. 3, which was obtained from the raw data shown in Fig. 2. The scattering data exhibit two pronounced peaks, which indicates that the film possesses structural order. This type of scattering behaviour is found irrespective of the actual D_2O-content of the films - as long as this content is below ~ 10%. It is a priori not possible to derive the type of structure in an unequivocal way from such sets of scattering data. But, in the present case we know from electron microscopic pictures that the latex spheres are packed regularly. Thus, we show in the following how the observed scattering behaviour can be explained by assuming that the latex spheres are close-packed in extended domains with a face centered cubic (fcc) lattice structure. The contrast which is necessary in order to obtain the observed scattering behaviour is due to the difference between the scattering length of the polymeric material in the latex spheres, on the one hand, and the scattering lengths of components of the interstitial phase, on the other hand. The interstitial phase consists of material which is not miscible with the polymers, as, e.g., salt, surfactants, D_2O, etc., and possibly of micro-voids. Here, it must be stressed again that the uptake of D_2O (below ~ 10%) by the films does not lead to a qualitative

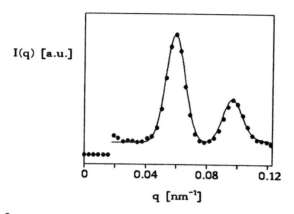

Fig. 3

Scattering data obtained from SANS on a latex film prepared at room temperature which contained ~ 2% D_2O. The six points on the left are points of zero intensity (beamstop). Solid line: see text.

change in the scattering curves. Hence, we can be sure that the assignment just given is correct. Making use of Babinet's theorem [12], we treat the data in a first order approach as if all scattering effects are solely due to the scattering by the latex particles. Problems arising with this approach will be described below.

Since the particles are fairly monodisperse we can write the scattered intensity I(q) as being proportional to the product of the particle form factor P(q) and the lattice structure factor S(q):

$$I(q) \sim P(q)\,S(q) \ . \tag{1}$$

The particle form factor which occurs because of intraparticle interference effects is given by [13]:

$$P(q) = 9\left(\left(\sin(qr) - qr\cos(qr)\right)/(qr)^3\right)^2 \ . \tag{2}$$

The structure factor of an ideal, infinitely extended lattice reads:

$$S(q) \sim \sum_{\{k_i\}} \delta(q - k_i) \tag{3}$$

$\delta(.)$ denotes the Dirac function. $\{k_i\}$ is the set of reciprocal lattice vectors (modulo 2π). When recording Debye-Scherrer diagrams (i.e. scattering by many randomly oriented crystallites) an averaging with respect to the azimuthal angle has been performed implicitly. Thus, only the modulus of q neglecting the vectorial character is of concern. In case of cubic symmetry the elements of the set $\{k_i\}$ are determined by

$$\left(2a/\lambda \right) \sin(\Theta_{hkl}/2) = (h^2 + k^2 + l^2)^{1/2} \tag{4}$$

where a is the lattice constant, h,k,l are the Miller indices. The k_{hkl} are given by $k_{hkl} = 4\pi \sin(\Theta_{hkl}/2) / \lambda$. Relations for $\{k_i\}$ for the case of non-cubic lattices can be found in any book on scattering methods as, e.g., Refs. [14,19]. The existence of certain (hkl)-peaks is ruled by extinction rules [14]. In the case of an fcc lattice there is constructive interference and thus scattered intensity if all h,k,l are even or all h,k,l are uneven ((111), (200), (220), (311), (222), ...). The corresponding Bragg peaks occur with the same intensity for all reflecting planes. The values of q for which peaks are to be found are listed in Table 1. In addition, the Bragg peak positions are given for the case that the fcc lattice is built by close packed spheres of diameter d=130nm which is the approximate average diameter of the latex spheres used. The fcc lattice constant is a=d√2. The values of the particle form factor P(q) at the positions of the peaks are also given in Table 1.

In Fig. 4 the contributions P(q) and S(q) to the scattered intensity

Index	(111)	(200)	(220)	(311)	(222)	(400)	(331)	(420)
$q \cdot a$ [1]	10.9	12.6	17.8	20.8	21.8	25.1	27.4	28.1
q [10^{-2}nm^{-1}]	5.9	6.8	9.7	11.3	11.8	13.6	14.9	15.3
$P(q)$ [a.u.]	1	$4 \cdot 10^{-3}$	0.4	$3 \cdot 10^{-2}$	$2 \cdot 10^{-4}$	$8 \cdot 10^{-2}$	$6 \cdot 10^{-2}$	$4 \cdot 10^{-2}$

Table 1

Bragg peak positions for an fcc-lattice with lattice constant a (first row) and a=184 nm = 130 $\sqrt{2}$ nm (second row). Third row: P(q)-values at the peak positions for spheres with diameter d=130nm.

I(q) are depicted for the case of an ideal fcc lattice built by monodisperse homogeneous spheres of diameter d=130nm. In Fig. 4 the intensity which is predicted for the SANS experiment on the ideal system described is indicated by the bold part of the vertical lines. Using the [P(q)S(q)] values thus determined and taking into account peak broadening effects one obtains the curve shown in Fig. 3. The coincidence with the SANS data is good. The observed broadening of the Bragg peaks is caused by several facts as will be discussed below. In a zero order approximation we describe the experimentally observed scattering curves by a convolution of the scattering contributions from an ideal fcc lattice consisting of close packed homogeneous spheres with a Gaussian taking up all effects which lead to the peak broadening. As fit parameters we have used the variance of the Gauss-function, describing the widths of both peaks, the scaling of the intensity, and a term for the flat background. The sphere diameter d=130nm and the lattice constant a=d$\sqrt{2}$ were used. Obvious deficiencies of this approach will be discussed below.

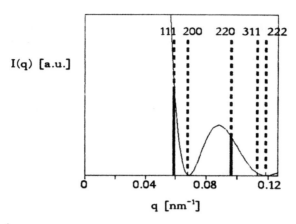

Fig. 4

Thin line: q-dependence of the particle form factor P(q) of a homogeneous sphere with diameter 130 nm. Bold lines (dashed and solid part): positions of the Bragg peaks S(q) as expected in the case of an ideal face centered cubic lattice made from close packed spheres of diameter 130 nm. The dashed parts indicate the degree of suppression of these peaks when calculating the total scattered intensity $I(q) \sim P(q)S(q)$: only two peaks are left (solid bold lines).

The crucial point for the understanding of the scattering data is the fact that three of the five expected Bragg peaks are strongly suppressed because of intraparticle interference effects which give rise to the multiplicative P(q)-term in relation (1). It is interesting to note that a similar effect has been recently reported for the case of close packed fullerites (spherical C_{60}-molecules) [15].

In order to check the good coincidence between the measured and the theoretically predicted data an additional SANS experiment has been performed at the D 17 instrument where a different q-range can be scanned ($0.05 \, nm^{-1} < q < 0.3 \, nm^{-1}$). The agreement between the resulting data and the predicted scattering behaviour of our model is good as regards the peak positions. It must be noted that the significance of the data is smaller for larger q-values because of the increasing efficiency of the P(q)-contribution which leads to a strong suppression of higher order peaks.

The lattice structure. So far we have found a good description of our data with the assumption of an fcc-lattice of close packed latex particles. In the following we discuss shortly whether modes of packing which are different from the fcc-lattice can be used for a description of the SANS data. Besides fcc packing, also hexagonal close packing (hcp) as well as the body centered cubic (bcc) lattice structure must be taken into consideration for the following reasons: 1. there are two types of organizing hard spheres into the closest possible packing: fcc and hcp [16], see also Ref. [17]. 2. It depends on the electrolyte concentration in the serum, on the surface charge density of the latex spheres, and on the temperature whether the

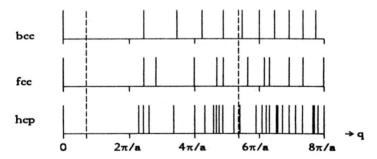

Fig. 5

Bragg peak positions of lattices built by close packed spheres with diameter d. The lattice constant a(d) is given in the text. The dashed lines indicate the range which is relevant for the discussion of the experiments presented here, see Fig. 3.

spheres order into an fcc or a bcc structure when approaching the state of touching each other during the film formation process [3,18]. In Fig. 5 the Bragg peak positions as expected for bcc, fcc, and hcp lattices made of close packed spheres with diameter d are depicted. Only the S(q)-contribution is shown. The q-range which was accessible in our experiments where d=130nm is enclosed by the dashed lines. It must be noted that the lattice constants are different in the three cases (bcc: $a=2d/\sqrt{3}$, fcc: $a=d\sqrt{2}$, hcp: $a=d$, $c=d\sqrt{(8/3)}$). The rules leading to the peaks shown in Fig. 5 are given in, e.g., Ref. 19. When comparing Figs. 5, 4, and 3 it becomes evident that only in the case of the fcc lattice made of close packed spheres there is sufficient agreement between the predicted and the observed scattering behaviour when taking into account the P(q)-term and the peak broadening effects. The subtle interplay between P(q)- and S(q)-contributions does not lead to a two-peak scattering profile when assuming bcc- or hcp-packing. One might argue that the P(q)-term as given in Eq. (2) cannot not be used since in the film the latex spheres are deformed into polyhedra (Fig. 1c). We checked this point by generating the P(q)-curves for spheres which were deformed to various degrees into polyhedra. A Monte Carlo method similar to the one described in Ref. [20] was used. With the use of these data and relation (1) it could be shown that only in the fcc-case there is good agreement between the predicted scattering behaviour and the SANS-data [21]. In addition, it was found that there is only a minor difference between the scattering curves for close packed spheres (volume fraction $\varphi = 71.1\%$) and close packed spheres which were compressed and deformed in such a way that $\varphi = 91.5\%$, see Fig. 6.

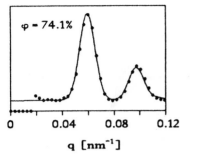 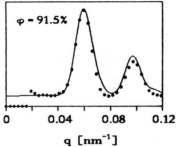

Fig. 6

Solid line: scattering of spheres close packed in an fcc-lattice (φ = 74.1%) and of fcc close packed spheres which are deformed into polyhedra such that the volume fraction is φ = 91.5%.

The contrast problem. As has been described above the data were treated in a simplified approach as if being due to a simple two-phase structure of the latex films with the corresponding two differing scattering lengths. We assumed that one phase is given by the polymers of the latex particles whereas the second phase, which is made from the non-polymeric material (salt, emulsifiers, etc.) and possibly micro-voids, was assumed to be homogeneous, see Fig. 7a. In a refined approach one should take into account that this second phase might possess an internal structure as is sketched in Fig. 7b. Very little is known about this internal structure. At present we have no choice but to continue the data analysis with the simple two-phase model. This approach is to a certain extent justified since the details of the interstitial structure will express themselves in scattering effects at q-values above the range which is of interest here - as long as the structures of the individual intersticies are not correlated.

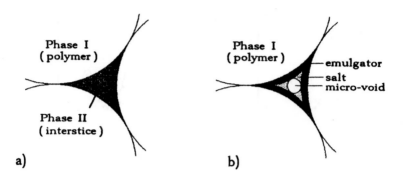

Fig. 7

Representation of the intersticies between the close packed polymeric latex particles. a) Simplified two-phase approach. b) Possible fine structure in an interstice.

On the quantitative description of the data. In the following we briefly mention the points which must be respected when trying to perform a fit of the data with a description going beyond the one presented here. A comprehensive discussion will be given elsewhere [22].

Apparatus effects: the wavelength distribution, the collimation effects, and the finite size of the detector elements lead to a broadening of Bragg peaks [23]. In addition, a q-dependent weakening of the scattering intensity is expected [23]. First attempts to describe our data with the theory proposed in Ref. [23] show that the width of the Bragg peaks, see, e.g., Fig. 3, is to a large extent due to these apparatus effects. Thus, the deconvoluted SANS-data should exhibit two sharp Bragg peaks indicating the considerable extension of the crystalline domains.

Sample effects: the width of the peaks is influenced by the finite size of the crystalline domains, by lattice defects, and by stacking faults [24]. The ratio of the intensities of the two peaks is determined by the following effects: multiplicities of lattice planes (in Debye-Scherrer diagrams), prefered deposition of latex spheres in (111)-planes along the film plane [4], lattice distortions, slight displacement of the particles from the vertices of the lattice, deformation of the latex spheres into polyhedra [21], and the values used for the size of the latex particle and for the actual lattice constant. As can be seen from this list, a complete description of the data is far from being trivial.

4. Further results

Film formation at elevated temperatures (70^0C). In Fig. 8 we compare the SANS-data obtained on films which were prepared at room temperature and at 70^0C, respectively. Four differences between the two sets of data are clearly discernible: the film which was prepared at 70^0C yields peaks with a larger width, the peaks are shifted to somewhat smaller q-values, there is an upturn in the scattered intensity at very small q-values, and the ratios between the two peaks differ between the two films.

These effects might be explained as follows: at elevated temperatures the water evaporates at a higher rate during the process of film formation thus leaving less time for the particles to arrange in a crystalline order. Therefore, the crystalline domains are to be expected of smaller size (larger width of the peaks). Possibly, the degree of lattice distortion is larger and there is less preferential deposition in (111) planes along the substrate (shift of the peaks, change of the

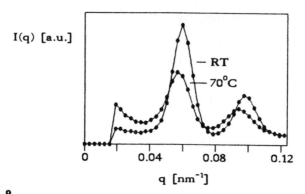

Fig. 8

SANS by latex films prepared at room temperature (RT) and at 70^0 C, respectively. The D_2O-content is ~6% in both cases. Drawn lines are guide lines for the eye.

ratio of the peak intensities). The amount of disordered regions (amorphous packing, domain boundaries, etc.) is larger. D_2O finds place in these regions to form easily aggregates of substantial size (scattering contribution at very small angles).

Tempering effects. When annealing the films at different temperatures above the glass transition temperature one observes two interesting effects. First, as is shown in Fg. 9a there is increasing coherent scattering of the <u>dry</u> films when raising the annealing temperature T_a. This might be explained as follows: upon tempering at elevated temperatures, material which is not compatible with the latex polymers, e.g., surfactants which are initially covering the surfaces of the latex particles, is expelled into the intersticies. Fig. 7b shows the SANS behaviour of films annealed at various temperatures with subsequent uptake of D_2O. For $T_a \leq 150^0C$ one observes the familiar two-peak structure which indicates that the water diffuses preferentially into the intersticies. The picture drastically changes once the film

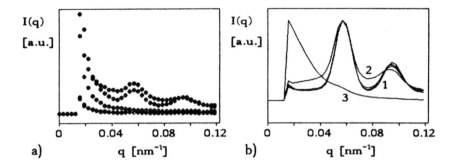

Fig. 9

SANS data obtained on films which were annealed for two hours at different temperatures T_a. a) Dry films; bottom to top: T_a= 120, 70, 150, 180°C. b) D_2O-content: 5%; T_a=70, 100, 120 (data 1), 150 (data 2), 180°C (data 3). In b) the intensities have been normalized to the maximal value of each data set.

has been tempered for two hours at 180°C: instead of the two peaks, which have almost completely disappeared, one observes an increasing SANS signal at very small q-values. We interpret this effect as arising from the existence of aggregates of water: since all the tempered films were prepared at room temperature thus possessing essentially the same morphology we conclude that the water no longer has access to the intersticies for $T_a \geq 180$°C and is thus forced to aggregate in other locations of the film. This, in turn, might be explained by the assumption that the percolation structure of the interstice network is lost when tempering for a sufficient time at $T_a \geq 180$°C. This effect is due to a relaxation and further deformation of the close packed particles and is accompanied by a decrease in the volume of the sample. Since the water is likely to diffuse along the percolation network of the intersticies, a loss of a certain fraction of bonds in this network makes part of the intersticies inaccessible to water. These points will be further examined in future experiments.

5. Summary

It has been shown that in the latex films which were examined the latex particles order in a polydomain face centered cubic lattice. The domain size decreases with increasing temperature of film formation. Water (D_2O) brought into the film diffuses into the intersticies which consist of non-polymeric material, as, e.g., salt, surfactants, and - probably - micro-voids. When annealing dry films at elevated temperatures additional non-polymeric material is expelled into the intersticies. These intersticies keep their identity even after prolonged tempering.

Ackowledgment

We thank Dr. A. Rennie for an interesting comment during a presentation of these results.

References

1 Luck, W., Klier, M., Wesslau, H. (1963). Ber. Bunsenges. 67, 75.
2 Pusey, P., van Megen, W. (1986). Nature 320, 340.
3 Robbins M.O., Kremer K., Grest G.S. (1988).
   J. Chem. Phys. 88, 3286.
4 Joanicot, M., Wong, K., Maquet, J., Chevalier, Y., Pichot, C.,
   Graillat, C., Lindner, P., Rios, L., Cabane, B. (1990).
   Prog. Colloid Polymer Sci. 81, 175.
5 Lissant, K.J. (1966). J. Coll. Interface Sci. 22, 462.
6 Geguzin, J.E. (1973). Physik des Sinterns, VEB Deutscher
   Verlag für Grundstoffindustrie, Leipzig.
7 Kast, H. (1985). Makromol. Chem. Suppl. 10/11, 447.
8 Hahn, K., Ley, G., Schuller, H., Oberthür, R. (1986).
   Colloid & Polymer Sci. 264, 1092.
9 Hahn, K., Ley, G., Oberthür, R. (1988).
   Colloid & Polymer Sci. 266, 631.
10 Linné, M.A., Klein, A., Miller, G.A., Sperling, L.H.,
   Wignall, G.D. (1988). J. Macromol. Sci.-Phys. B27, 217.
11 Ghosh, R.E. (1989).A Computing Guide for Small Angle Scattering
   Experiments, ILL, Grenoble.
12 Guinier, A., Fournet, G. (1955). Small-Angle Scattering of X-Rays,
   Wiley, New York.
13 Hunter, R.J. (1987). Foundations of Colloid Science, Vol. I,
   Clarendon Press, Oxford.
14 Azároff, L.V., Bueger, M.J. (1958). The Powder Method,
   McGraw-Hill, New York.
15 Fischer, J.E., Heiney, P.A., McGhie, A.R., Romanow, W.J.,
   Denenstein, A.M., McCauley, J.P., Smith, A.B. (1991).
   Science 252, 1288.

16 Atkins P.W. (1983). Physical Chemistry, Oxford University Press, Oxford.

17 Max, N. (1992). Nature 355, 115.

18 Monovoukas, Y., Gast, A.P. (1989). J. Coll. Interf. Sci. 128, 533.

19 Guinier A. (1956). Théorie et Technique de la Radio-crystallographie, Dunod, Paris.

20 Hansen, S. (1990). J. Appl. cryst. 23, 344.

21 Dippel, O., Rieger, J. (1992). Unpublished results.

22 Rieger, J., Hädicke, E., Ley, G., Lindner, P. (1992). To be published.

23 Pedersen, J.S., Posselt, D., Mortensen, K. (1990). J. Appl. Cryst. 23, 231.

24 Kleeberger L., Ruppersberg H. (1977). Z. Metallkunde 68, 742.

AN EFFECTIVE HARD-SPHERE MODEL OF THE NON-NEWTONIAN VISCOSITY OF SOFT-SPHERE DISPERSIONS

Richard Buscall

ICI Corporate Colloid Science Group
P.O. Box 11, The Heath, Runcorn
Cheshire WA7 4QE, United Kingdom

Of all of the readily-measured, physical properties of colloidal dispersions the non-Newtonian viscosity is perhaps the most sensitive to the nature of the interactions between the particles. An empirical correlation for the viscosity of soft-spheres proposed recently has been compared with experimental data for small, sterically-stabilised latices. The model relates the non-Newtonian viscosity to an apparent or notional pairwise interaction-potential. It is shown to scale the available data remarkably well.

INTRODUCTION

Dispersions of fine particles show a wide range of rheological behaviour. In crude terms they can vary from solid pastes to thin fluids. The precise behaviour seen depends in part upon the solids content but it is mediated by particle interactions to the point where, in the sub-micron particle-size regime, the whole range of possible behaviour can often be elicited from the same basic dispersion by manipulating the physico-chemical variables that influence the interactions. Because of this wide range of possible behaviour it would clearly be useful to be able to predict the rheology of such disperse systems and the effect of interactions between components upon it. At present it is not possible to do this except at a *very* qualitative level.

The work described in this paper is concerned with the effect of repulsive interparticle forces (i.e. the forces giving colloidal stability) on the rheology of sub-micron dispersions. When the work was first embarked upon aim was to develop an approximate empirical correlation for the electroviscous effect in aqueous dispersions, with the idea of removing the ambiguity the latter can cause when viscosity measurements are used to assess the stability of concentrated systems. However, it soon became clear from preliminary comparisons with experimental data that the approach developed for this limited purpose had a greater potential. An obvious area of application was that of ultrasmall latices, microgels and self-dispersible resins.

The model is validated using data for two ultrafine latices developed by ICI - an NAD [1] of 84nm mean particle-diameter, and an aqueous latex (PLANIGALE - planigale is the smallest Australian marsupial) of somewhat similar size, both latices being sterically-stabilised.

EFFECTIVE HARD-SPHERE MODEL

The essential features of the model are summarised here, more detail and explanation is given in reference 2. A point worth making at the outset is that the model is very simple. It boils down to a couple of algebraic equations, and as such it is very easy to use.

Bearing in mind that the *viscosity* of a system is defined as the ratio of the *stress* to the *shear-rate* in steady flow, and that either of the latter two variables might be chosen as the independent variable, the flow behaviour of a system can be described in terms of the relationship between any two of the three possible variables. That is, a knowledge of any one of the following, as a function of the variable in brackets suffices to characterise the flow,

viscosity (stress)
viscosity (shear-rate)
stress (shear-rate)

This rather obvious point is laboured because the shear-rate, commonly the independent-variable in experimental practice, will not be referred to explicitly in the following.

Now to the essentials of the approach. There are recognized to be four contributions to the stress required to maintain flow in a solution or dispersion [3]:-

$$stress = \left\{ \begin{array}{l} thermodynamic \ + \\ thermal\,(Brownian) \ + \\ hydrodynamic \ + \\ inertial \end{array} \right\}$$

It can be argued with some justification that the rheology will only be "interesting", i.e. *significantly* different from that of the dispersion medium, when the first contribution, for which read "thermodynamic = particle-interactions", dominates the others [3]. This is an attractive notion because it suggests the possibility of a unique and direct relationship between the bulk flow behaviour and the microscopic force-law (or potential) characterizing the particles and their interactions; it points to a very direct connection between the flow and the colloid chemistry.

The model presented here is a guess of what this relationship might look like. It should be emphasised that it is no more than a guess, backed up by dimensional analysis. Its sole justification is its capacity to account for (scale) real experimental data successfully.

The model embodies the idea that colloidal particles interacting via soft repulsive potential have an effective, shear-dependent, collision diameter R_{eff}. Further, that the viscosity of a suspension of such particles is equal to the viscosity of a fictitious dispersion of hard-sphere-like particles of size R_{eff} at the same number concentration. These assumptions then allow the viscosity to be estimated from known correlations for hard-spheres (e.g. eqtn. 1 below). The interparticle pair potential $U(R)$ and the stress (σ), enter by equating the potential at the collision diameter $U(R_{eff})$ to the mean flow energy-density per particle, $\sim \sigma R_{eff}^3$, again the precise form used is given below (eqtn. 3). It is here where the guess-work comes in.

In the absence of any knowledge of the microscopic kinematics in a concentrated dispersion it is difficult to postulate a prescription for the collision-diameter. Equation 3 is an ad-hoc guess.

The viscosity η is related first to an effective solids concentration, expressed as a volume-fraction ϕ_{eff}, by the Dougherty-Kreiger equation [4],

$$\frac{\eta}{\eta_s} \cong \left(1 - \frac{\phi_{eff}}{\phi_{max}}\right)^{-\frac{5}{2}\phi_{max}} \qquad [1]$$

This is one of a number of correlations that might have been chosen, its significance is that it is known to account for the concentration dependence of hard-sphere like particles rather well [5].
From the effective volume-fraction the collision diameter R_{eff} is obtained through the obvious metric identity,

$$\phi_{eff} = \phi\left(\frac{R_{eff}}{2a}\right)^3 \qquad [2]$$

where ϕ is the true volumetric solids-content and a is the true mean particle radius (the latter may not be known, or in the case of microgels may have no meaning, but assume that it is and does for the moment).

The third, key equation connects the interaction potential $U(R)$ to the shear stress σ and reads [2],

$$\frac{U(R_{eff})}{k_B T} = \frac{1}{2}\left[\frac{\sigma \phi_{eff} a^3}{\phi K(\phi_{eff}) k_B T} + 1\right] \qquad [3]$$

where [2], $K(\phi)$ is the dimensionless characteristic stress for hard-spheres. The experimental data available [6] show $K(\phi)$ to have an approximately linear dependence on volume-fraction within appreciable scatter, linear regression to this data gives,

$$K(\phi_{eff}) = 0.016 + 0.52\phi_{eff} \tag{4}$$

Equations 1 to 4 serve to connect all the relevant variables. Thus, given $U(R)$ the viscosity as a function of shear stress (or rate) can be calculated.

In this form the model is not particularly useful *per se* since the potential will not be known in any real example unless it is very contrived. Any utility the model has stems from the possibility of using it *first* in reverse to obtain an apparent potential from a flow-curve, and then using the latter as input to make predictions of the effect of varying, say, particle-size, or concentration, or whatever. In addition, the use of the model to obtain an apparent potential can have analytical implications since the inter-particle potential contains information on the surface structure (and on particle-size, clearly). For example, for sterically stabilised particles and microgels (i.e. where the interaction is "osmotic") the potential is related to the radial segment-density distribution of the outer, solvated layer.

The model has been validated in two ways [2]; by comparing experimental viscosity data for simple ionic latices with predictions made using theoretical potentials, and, by seeing how well it correlates (collapses or reduces) data for sterically-stabilised latices.

The results for simple, ionic latices are discussed in reference 2, as are some of the available results for model sterically-stabilised latices. However, because of the importance of sterically-stabilised systems, and because some new data has become available since the writing of [2], this aspect is considered from scratch below.

RESULTS & DISCUSSION

In order to challenge the model data are needed for otherwise well-characterised particles for which the thickness of the solvated stabilising-layer is appreciable relative to the size of the core, i.e. the particles have to be small and/or the layer thick. Only two suitable sets of data have been found, although more data is expected to become available shortly [7].

The results for non-aqueous latex (NAD) will be considered first. The experimental data used here are taken from the thesis of Frith [8]. The NAD used was a dispersion of PMMA in decalin, stabilised by poly(12-hydroxystearic acid) - PHS, made according to the basic procedures developed by ICI Paints [1]. The mean particle diameter was 84nm by electron microscopy.

Flow curves are shown in fig. 1, the viscosity and shear stress being given in reduced units, viz.,

$$\eta_R = \frac{\eta}{\eta_{decalin}}$$

$$\sigma_R = \frac{\sigma a^3}{k_B T}$$

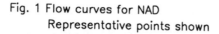

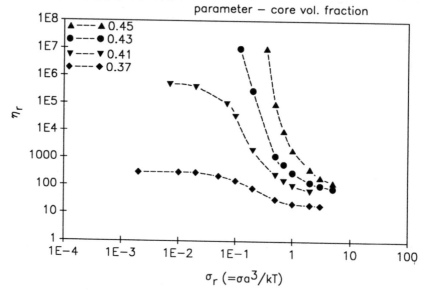

Fig. 1 Flow curves for NAD
Representative points shown
parameter — core vol. fraction

Fig.1 Log-log plot of relative viscosity versus reduced shear stress for NAD. Data taken from Frith [8]. The parameter is the volume-fraction of the PMMA core. Only representative data points are shown, and two curves at intermediate concentrations of 0.402 and 0.389 have been ommitted. For the calculations which follow, fits of the full data set to the Williamson equation (eqtn.7) given by Frith [8] were used.

The parameter differentiating the various curves is the volume-fraction of PMMA, which ranges from 0.37 to 0.45. Notice that the flow curve changes considerably over this fairly narrow concentration range. At ϕ_{PMMA}=0.37 rather gentle shear-thinning is seen with a ten-fold drop in viscosity over two decades in stress. At ϕ_{PMMA}=0.41 the lower plateau is lost and the dispersion shows a yield stress. Overall, something like a million-fold change in viscosity can be seen. This behaviour is attributed to interpenetration of the solvated PHS layers.

The model can be used to obtain an apparent interaction potential from any one flow curve. Thus a test of the model is to see whether different flow curves yield the same potential. This is a powerful test given the order-of-magnitude or more difference in low-shear viscosity between each successive curve. In his thesis Frith shows flow curves at six different volume-fractions. Calculations were performed for the four volume-fractions for which Frith provided curve-fits to the data, the data not being readily available in tabular form. It should be said however that the other two flow curves are entirely consistent with the picture that emerges using these four [2]. The results are shown in fig. 2, where the potential is plotted against the distance of closest approach of the PMMA cores,

$h = R - 2a$

so as to show the detail.

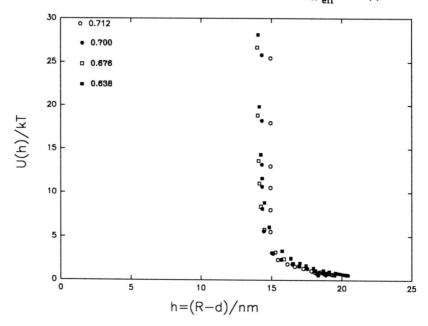

Fig. 2 Interaction potential from viscosity for NAD
parameter:− effective volume fraction ($\varphi_{eff}=1.74\varphi$)

Fig.2 Apparent interaction potential for NAD derived from flow curves
using the EHS model. The parameter distinguishing the curves is the
apparent volume-fraction quoted by Frith, this being obtained by
multiplying the PMMA core volume-fraction by 1.74.

The first comment to make is that the potential looks sensible. It has a soft-tail and a steep core, as might be expected given that the stabiliser is of low MW but polydisperse, and the surface coverage is high [1,9]. The distance scale also looks correct - the effective thickness of the PHS layer on model NAD have been measured by several methods and several groups, the methods including intrinsic viscosity, PCS and SANS. Mean thicknesses of 9 ± 1nm have been obtained, typically [1,8], from which significant interaction would be expected at $h = 18 \pm 2$nm. This figure lies nicely between the value of h where the interaction starts (~21nm) and that where it hardens (~15nm).

The potentials from three of the flow-curves superpose to better than 1 nm on the abscissa. This should be compared with R~100nm not h~15nm. The potential from the fourth curve deviates somewhat at close-approach. However, inspection of the original figure in Frith's thesis [8] shows that in this case the curve fit did not follow the data too well, and this is the cause of the difference. Thus, the model has reduced four curves spread over four decades to a single curve, about which the scatter is less than 1%. There can be no arguing that the model scales the experimental data correctly, even though it may not be clear why the model works so well.

In principle the potential contains information on the density distribution in the surface layer, the latter is however not at all easy to extract since the form of this dependence is like [9],

$$\int_{h}^{\infty} U(h')\mathrm{d}h' \sim \int_{0}^{h} \rho(x)\rho(h-x)\mathrm{d}x \qquad [5]$$

The data for an aqueous system are now considered. The data for
Planigale latex used were made available by Dr. D. A. R. Jones of
Melbourne University and B. Leary of Dulux (Australia). The Planigale
latex is a small aqueous, acrylic latex made using a non-ionic surfactant
with a mean degree of ethoxylation of ca. 10. The particle diameter was
quoted as 58 ± 12nm. The solids content of the samples was given on a
mass basis and the volume-fractions of the acrylic core were not known.
However, the intrinsic viscosity of the latex had been measured and so an
apparent vol.fraction was calculated from,

$$\phi_{app} = \frac{2}{5}[\eta]c \qquad\qquad [6]$$

where the intrinsic viscosity is in dl/gm and c is the weight % solids. This
measure of the volume-fraction contains contributions from both the core
and the surfactant-layer.

The flow curves are shown in figs. 3 and 4, the parameter attached to the
curves being the apparent volume-fraction.

Fig. 3 Flow Curves For Planigale

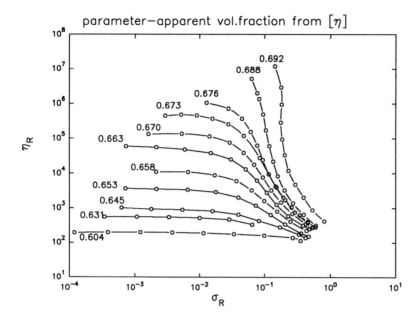

RHEOL/PLNGLFC..SPG

Fig.3 Flow curves for Planigale latex. The parameter distinguishing the curves is the apparent volume-fraction calculated from equation 6 (data courtesy of Boger, Jones & Leary).

Fig. 4 Flow Curve for Planigale at $\varphi_{\text{eff}}=0.670$

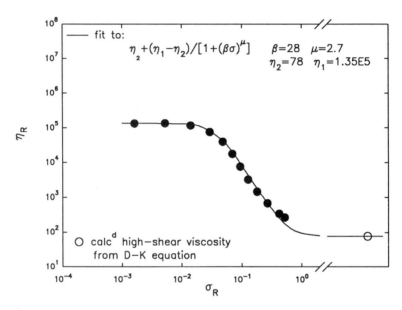

RHEOL/PLNGLFC..SPG

Fig.4 Flow curve for Planigale latex at an apparent volume-fraction of 0.67.

The curves are similar, qualitatively, to those for NAD. The potentials derived from the curves are shown in fig. 5.

Fig. 5 Apparent Potential for Planigale

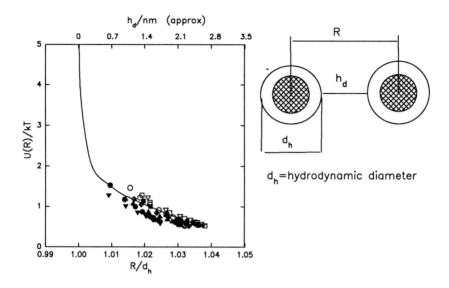

Fig. 5 Apparent interaction potential for Planigale derived from the flow curves shown in figs. 3 & 4.

Again the scaling of the data is excellent. In this case only the tail of the potential is seen and not the steep rise. This is because the flow curves do not extend to high enough stresses to show an approach to a high-shear plateau (the slope of the apparent potential bears an inverse relation to the gradient of viscosity-stress curve). In one case, what might or might not be thought a plausible extrapolation of the flow curve was made (fig.4). The extrapolation was constructed by fitting the Williamson equation,

$$\eta(\sigma) = \eta_2 + \frac{\eta_1 - \eta_2}{1 + (\beta\sigma)^\mu}$$ [7]

to the data using β and μ as adjustable constants. The other unknown
parameter η_2 was constrained by choosing a value which gave from the
Dougherty-Krieger equation, equation 1, a value for the effective
volume-fraction at infinite-stress which coincided with the apparent
volume-fraction calculated from the solids weight-concentration and
intrinsic viscosity using equation 6. The potential obtained from the
extrapolated curve is shown as the solid line. Different choices for η_2 turn
out to make rather little difference to the form of the predicted potential
provided they represent sensible extrapolations of the data in fig.4.

CONCLUSIONS

A empirical correlation for the viscosity of soft-spheres proposed
recently has been compared with experimental data for small,
sterically-stabilised latices. The model relates the non-Newtonian
viscosity to an apparent or notional pairwise interaction-potential. It is
shown to scale the available data remarkably well. The model has also
been applied to the data of Wolfe and Scopazzi [10] for model, swellable
microgels, with similar success. This work will be discussed elsewhere.

Acknowledgement

D. Boger and D.A.R. Jones (Chemical Engineering Dept., Melbourne University) and B. Leary (Dulux (Australia)) are thanked for providing the Planigale data.

REFERENCES

1] K.E.J. Barrett, "Dispersion polymerisation in organic media", J. Wiley & Sons, London (1975)

2] R.Buscall, J. Chem. Soc. Faraday Trans., **87**(9) 1365 (1991)

3] W.B. Russel, D.A. Saville & W.R. Schowalter, in "Colloidal dispersions", Cambridge University Press, (1989), chapter 14.

4] I.M. Kreiger, Adv. Colloid Interface Sci., **3** 111 (1972)

5] e.g. R. Buscall, Faraday Discuss. Chem. Soc., **76** 338 (1983)

6] Ref. 3, fig. 14.4(b), p. 468

7] P. d'Haene & J. Mewis, K.U.Leuven, private communication.

8] W.J. Frith, Doctoral thesis, Katholieke Universiteit Leuven (1986)

9] D.H. Napper, "Polymeric stabilisation of colloidal dispersions", Academic Press (1983)

10] M.S. Wolfe & C. Scopazzi, J. Colloid Interface Sci., **133** 265 (1989)

Oscillatory Flow of Polymerically Stabilised Suspensions

P. D'Haene and J. Mewis

Chem. Eng. Dept., K.U.Leuven, 3001 Heverlee,Belgium

April 26, 1992

Abstract

The dynamic moduli of a number of sterically stabilised, monodisperse, poly(methylmethacrylate) latices have been measured to complement earlier data on similar systems. The results are used to test the validity limits of Brownian hard sphere scaling, to evaluate the effect of stabilizer layer softness and to correlate the data with fundamental colloidal parameters of the system under investigation. A strong similarity is found between the softness effect on steady state viscosity and that on the storage moduli. The shape of the relaxation spectra changes with concentration in a way which is identical to that of the zero shear viscosity and comparable with available hard sphere results. Attempts to compute the moduli from a perturbation theory did not result in a good relation between storage moduli and volume fraction. Using a simple lattice model, following Buscall, resulted in an excellent data reduction, from which an interaction potential was calculated for the PMMA latices. An effective hard sphere diameter, calculated from this potential, adequately reduces the data.

1   Introduction

Concentrated, stable, colloidal suspensions display viscoelastic behaviour. This can be probed directly in oscillatory flow, where the storage modulus (G') and the loss modulus (G'') measure respectively the

elastic and the viscous contribution. The elastic effects can be caused by two different mechanisms. The first is provided by Brownian
motion and can occur in all stable colloids, it is the only possible mechanism in Brownian hard spheres [1]. In sterically, i.e. polymerically,
stabilised systems the interparticle potential, caused by the deformation of the stabiliser layer, can also contribute to the elasticity.

Experimental results for dynamic moduli of hard spheres are available [2] and can serve as a reference, even if the theory is not totally
established yet [3]. A number of experimental results are available
for polymerically stabilised systems [4, 5, 6, 7]. Very few satisfy the
normal theoretical requirements of being within the linear amplitude
region, covering a substantial frequency range and having the relevant parameters changes systematically changed. Here, linear data
over a wide range of frequencies are presented with concentration and
particle size as variables. They supplement earlier data [6] on similar
systems. It is attempted to correlate the deformability of the stabiliser
layer with the viscoelastic properties of the suspension. In addition
scaling laws for the relaxation times are considered. The final goal of
this work is to find the basic relations between colloidal characteristics
and rheological properties in order to predict one from the other for
sterically stabilised dispersions.

2   Experimental

Extensive information about colloidal characteristics, rheological properties and structure is known for poly(methylmethacrylate) (PMMA)
particles which are stabilised by grafted chains of poly(hydroxystearic
acid) (PHS), e.g. in [8, 9, 10, 11, 6]. The samples were prepared according to the procedure described in [12]. The subsequent cleaning
and characterization has been described in [13], where also the steady
state viscosities of similar systems, but with different particle sizes,
have been discussed. The suspending medium is a mixture of cis-
and trans-decahydronaphtalene (decalin). From the measured values
of the intrinsic viscosity the thickness of the PHS layer is calculated
to be about 9 nm. Several particle sizes have been prepared. For
the larger ones, the linear region for the dynamic measurements be-

comes systematically smaller, making eventually linear measurements impossible. Dynamic measurements will be reported for particles with 129 and 376 nm core diameter (standard deviation on particles size respectively 13.5 and 6.1%).

The rheological measurements have been performed on a Rheometrics 705F. All data reported here are taken at 20.0°C. It is known that, around this temperature, the rheological data for different temperatures can be superimposed with the Brownian hard sphere scaling [13, 6]. Strain sweeps were performed to identify the limit for linear behaviour. The real measurements were then performed in the linear strain region. For the bigger particles it became impossible to obtain linear loss moduli at high volume fractions. The reported storage moduli are all in the linear region.

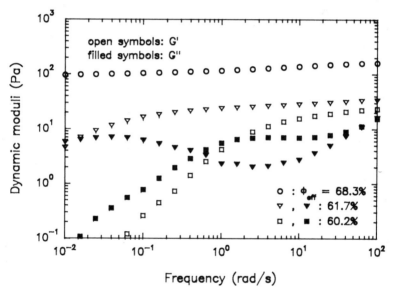

Figure 1: *Storage moduli G' and loss moduli G'' for concentrated suspensions of polymerically stabilised particles (diameter = 129 nm)*

3   Dynamic Moduli

The dynamic moduli for the more concentrated dispersions, containing 129nm particles, are shown in figure 1. The particle concentration is expressed as an effective volume fraction ϕ_{eff}. This includes the volume of the PMMA core and that of the stabiliser layer. The thickness of the latter is derived from viscosity measurements at small concentrations as indicated above. Hence the effective volume fraction expresses the unperturbed volume of the total particle. At high volume fractions and high shear rates the stabiliser layer will deform, making the effective volume a less meaningful parameter. For the smaller volume fractions of figure 1, and at lower frequencies, a typical fluid-like behaviour is observed. The loss moduli increase then proportional to the frequency and the storage moduli proportional to the square of the frequency. From the former a low shear Newtonian viscosity can be calculated. Within measuring accuracy it equals the corresponding value from steady state shear flow.

At the high frequency limit the loss moduli seem to tend to another Newtonian region. This can only be seen at the highest concentrations where no high shear plateau could be reached in steady state shear flow. The storage moduli display a plateau value at high frequencies. Except for the limiting viscosities, two fundamental characteristics can be derived from the dynamic data. The plateau of G' reflects an equilibrium, frequency-independent, elastic contribution. For Brownian hard spheres this is caused by a thermodynamic, Brownian, contribution. In the present case, plateau values start to be detected at volume fractions which approximately correspond to the maximum packing fraction at zero shear rate. The highest volume fractions that can be reached are considerably higher than the maximum possible packing for monodisperse hard spheres. Hence it is concluded that the stabiliser layer is gradually more compressed as the volume fraction is increased. The plateau moduli should then be governed primarily by the steric repulsion forces. The plateau moduli which have been measured here, together with those measured earlier [13], are shown in figure 2.

In order to compare with the expected result for Brownian hard

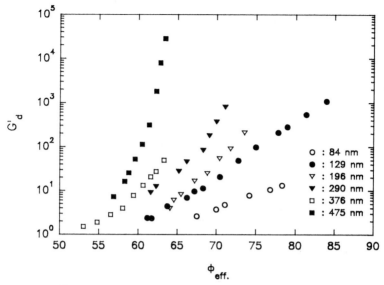

Figure 2: *Effect of particle size on the concentration dependency for the reduced plateau moduli*

spheres, the scaling for such spheres has been applied to the measurement results:

$$G'_d = \frac{G'a^3}{kT} \tag{1}$$

$$G''_d = \frac{G''a^3}{kT} \tag{2}$$

$$\omega_d = \frac{\eta_m \omega a^3}{kT} \tag{3}$$

where η_m is the medium viscosity, a the particle radius, T the absolute temperature and k the Boltzmann constant.

Using these dimensionless moduli, the data for the different particle sizes should superimpose. The dimensionless frequency is based on the diffusivity for dilute suspensions and should not reduce the data for different concentrations. It is obvious from figure 2 that the PMMA suspensions do not obey the hard sphere scaling, confirming the argument from the previous paragraph.

A theoretical framework for the high frequency moduli is given by the analysis of Zwanzig and Mountain [14]. It averages the contributions from potential (Φ) interactions between pairs of particles over the pair distribution function $g(r)$ for a particle number density ρ:

$$G'_{\infty} = \rho kT + \frac{2\pi}{15}\rho^2 \int_0^{\infty} g(r)\frac{d}{dr}\left(r^4\frac{d\Phi}{dr}\right) dr \tag{4}$$

Equation 4 was used with a perturbation model for g(r) and various expressions for the potential Φ to simulate the data of figure 2. The results are comparable with those of Goodwin and Ottewill [7], but the experimental data could not be reproduced adequately in this manner. As an alternative, the approach suggested by Buscall [15] has been used. It assumes a nearly hard sphere potential and a number of nearest neighbours N_1 on a fixed distance R_m, the mean separation distance, which depends in turn on the volume fraction and the maximum packing $\phi_{m,c}$ for the core particles.

From the relation between plateau modulus and potential a function $F(R_m)$ can be derived:

$$F(R_m) = \frac{d^2\Phi/dr^2}{D_c kT} \approx \frac{5\pi}{\phi_{m,c}N_1}\left(\frac{G'_{\infty}}{kT} - \frac{6\phi_c}{\pi D_c^3}\right) \tag{5}$$

In equation 5 the mean separation distance at a given core diameter D_c depends on the core volume ϕ_c and the corresponding maximum packing by:

$$R_m = D_c\left(\frac{\phi_{m,c}}{\phi_c}\right)^{1/3} \tag{6}$$

Assuming the Deryaguin approximation to convert the flat plate interaction potential to spheres, it follows that $F(R_m)$ should be independent of particle dimensions. In order to calculate $\phi_{m,c}$ and N_1 a random, glassy, arrangement has been assumed, consistent with the experimental value of the maximum packing for the limiting low shear viscosity. This leads to a single curve for $F(R_m)$ for all particle sizes of monodisperse model suspensions (figure 3). This is not the case if an fcc packing is assumed. The second curve on the figure refers to the diameters 196 nm and 290 nm, which were not completely monodisperse and therefore look somewhat softer. It should be mentioned that the figure is based on a maximum packing of 0.61 for the 475 nm particles. This value differs from the one reported originally (0.622) [6]. A comparison with the present modulus-concentration curves showed an inconsistency near maximum packing which is resolved by adjusting the volume fraction slightly.

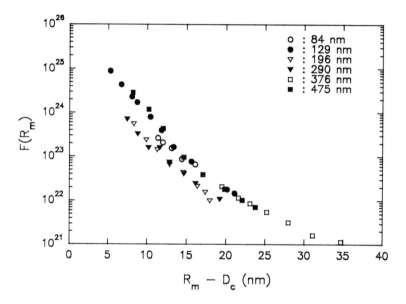

Figure 3: *Superposition of the data for the function $F(R_m)$ for various particle diameters*

Over the interparticles distances which can be covered, the potential can be represented by equation 7, if the interparticle distance r is expressed in nm:

$$\frac{\Phi(r)}{D_c kT} = 1.3 \cdot 10^{10}(r - D_c)^{-3.4} \tag{7}$$

The hydrodynamic thickness δ of the stabiliser layer is 9 nm. Yet the given procedure for calculating the interparticle potential gives non-zero values at much larger distances than 2δ. Various phenomena might be responsible for this. The presence of a molecular weight distribution for the stabiliser molecules is one of them. Alternatively there is the possibility of weak Coulombic repulsion [16]. Finally it is also possible that the nearest neighbour approach starts to fail at such large interparticle distances.

Russel [20] has suggested a scaling, also starting from the Zwanzig and Mountain theory, which uses the dimensionless group $G'_{\infty} a_c \delta^2 / kT$, which is essentially plotted as a function of the mean interparticle distance. For a constant value of δ this procedure gives similar results as the one used here, including the shape of the potential curve.

Polymerically stabilised particles are still relatively hard. Various features of their behaviour can be described by substituting the real interparticle potential by a hard sphere potential with a corresponding effective hard sphere diameter D_{eff}. Possible procedures for calculating D_{eff} have been reviewed by van Megen and Snook [17]. Here, the Barker-Henderson perturbation theory [18] has been followed. The resulting effective diameters will be used later to compare particle sizes. They can also be applied for reducing the viscosity curves.

4   Relaxation Times

The transition from fluid-like to solid-like behaviour occurs over a given frequency range. It can be characterized by the frequency at which viscous and elastic contributions are exactly in balance, i.e. where G' = G". This provides a measure for the dominating relaxation time of the slow material response. Eventual fast, high frequency, relaxations associated with relaxational modes of stabiliser molecules, are not considered here. The dominating relaxation time τ directly probes the particle mobility and hence is closely related to the particle diffusivity. Its change with particle size and concentration is shown in figure 4.

The extreme sensitivity of the relaxation time for the volume fraction is well known [13]. Here, also the pronounced effect of particle size is demonstrated. This is an expected result for Brownian hard

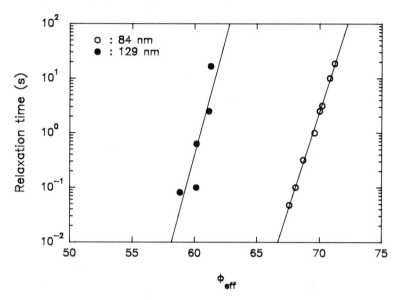

Figure 4: *Effect of particle size and concentration on the relaxation time (data for 84 nm particles from [6]).*

spheres, where the relaxation time changes with the third power of the radius, see equation 3. The line for the 84 nm particles is less steep than that for the 129 nm particles. This reflects the decrease in softness with increasing particle size. The two other sizes, which are not monodisperse, do not fit in with the monodisperse samples but are consistent with each other. The relaxation times are smaller and the curves less steep than expected on the basis of the average diameter. Particle size distribution evidently entails always an apparent softness.

The relaxation time provides a scaling factor for the frequency dependence. In figure 5 the curves for the loss moduli at different volume fractions are plotted versus the reduced frequency $\omega\tau$. This superimposes the limiting low shear moduli, and consequently also the limiting low shear viscosities. Furthermore, it shows that the basic relaxation spectrum is essentially identical for all concentrations, but it is increasingly cut off by the high frequency limiting viscosity at lower concentrations.

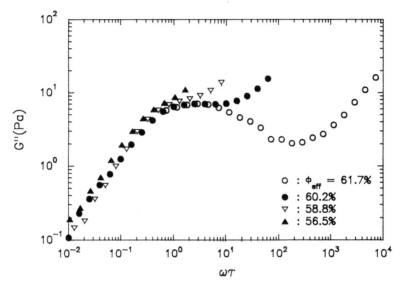

Figure 5: *Reduction of the frequency scale for the loss moduli by means of the relaxation time at different concentrations (diameter=129 nm)*

The particle size dependence of the relaxation time follows fairly closely that for hard spheres, i.e. $\propto a^3$. However, in figure 4 it has been seen that the concentration dependence differs slightly for the two particle sizes. The relaxation times express the particle mobility, hence they could be compared with the diffusional time scales. For the latter we use, in non-dilute systems, the zero shear viscosity η_0 instead of the medium viscosity: $6\pi\eta_0 a_{eff}^3/kT$. In this expression the effective hard sphere radius has been used for the particle size. The ratio between the two characteristic times is shown in fig 6.

From this figure it is obvious that the strong concentration effects in relaxation time and zero shear viscosity balance each other. Both express directly the particle mobility. Also the particle size effect is properly taken into account because the curves for the two particle sizes superimpose. A similar result has been obtained by van der Werff et al. [19] for Brownian hard spheres. Their data superimpose on the present ones for polymerically stabilised systems. However, they defined the relaxation time and the viscosity in a somewhat different

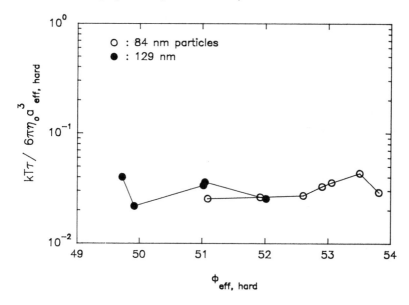

manner. Hence the absolute superposition with our data might be incidental.

5 Conclusions

Dynamic viscoelastic data are reported for a series of polymerically stabilised, monodisperse suspensions. Their behaviour deviates systematically from that of Brownian hard spheres. The limiting high frequency storage moduli for various particle sizes superimpose on a single curve if plotted against interparticle distance. From these data an interparticle potential can be calculated for each particle size. The potential in turn can be used to calculate an effective hard sphere diameter which describes the effect of particle size.

For a number of systems a characteristic relaxation frequency can be derived from the data. Its concentration dependence follows that of the limiting zero shear viscosity. The relaxation time can be used as a direct measure of particle self diffusivity. This clearly shows the close link between rheological properties and colloidal characteristics of sterically stabilised suspensions. As a result the rheological properties can be predicted from other measurements of the characteristics or the latter can be derived from rheological data.

Acknowledgement

This work has been supported by ICI, the International Fine Particle Research Institute and the Nationaal Fonds voor Wetenschappelijk Onderzoek (Belgium).

References

[1] Russel, W.B. and Gast A.P. (1986). J. Chem. Phys. 84, 815.

[2] van der Werff, J.C. et al. (1989). Phys. Rev. A 39, 795-807.

[3] Wagner, N.J. and Russel, W.B. (1989). Physica A 155, 475-518.

[4] Milkie, T. et al. (1982). Coll. Pol. Sci. 260, 531-535.

[5] Strivens, T.A., (1983). Coll. Pol. Sci. 261, 74-81.

[6] Frith, W.J. et al. (1990). J. Colloid Interface Sci. 139, 55-62.

[7] Goodwin, J.W. and Ottewill, R.H. (1991). J. Chem. Soc. Faraday Trans. 87, 357.

[8] Cebula, D.J., et al. (1983). J. Colloid Interface Sci. 261, 555.

[9] Pusey, P.N. and van Megen, W. (1987) in: *Physics of Complex and Supramolecular Fluids* (Eds. S.A. Safran and N.A. Clark), J. Wiley, New York.

[10] Livsey, I. and Ottewill, R.H. (1989). Colloid & Polymer Sci. 267, 421-428.

[11] Ackerson, B. (1990). J. Rheol. 34, 553-590.

[12] Antl, L. et al. (1986). Colloids and Surfaces 17, 67.

[13] Mewis, J. et al. (1989). AIChE J. 35, 415-422.

[14] Zwanzig, R. and Mountain, R.D. (1965). J. Chem. Phys. 43, 4464.

[15] Buscall, R. (1991). J. Chem. Soc. Faraday Trans. 87, 1365.

[16] Adriani, P.M. and Gast, A.P. (1990). Faraday Disc. Chem. Soc., 90, 17.

[17] van Megen, W. and Snook, I. (1984). J. Colloid Interface Sci. 100, 359.

[18] Barker, J.A. and Henderson, D. (1967). J. Chem. Phys. 47, 4714.

[19] vand der Werff, J.C. et al., (1989). Phys. Rev. A 39, 795.

[20] Russel, W.B. (1991). MRS Bulletin 16(8), 27.

The Preparation and Characterization of PolyNIPAM Latexes.

Robert Pelton*, Xiao Wu, Wayne McPhee and Kan Chiu Tam

Department of Chemical Engineering
McMaster University
Hamilton, Canada, L8S 4L7

Abstract:

Reviewed is the preparation and characterization of water swollen colloidal microgel particles prepared from cross-linked poly(N-isopropylacrylamide), polyNIPAM. Swelling, rheological and electrokinetic properties are presented. All properties of the latexes showed extreme temperature sensitivity near the cloud point temperature of polyNIPAM in water.

Introduction:

Water based polymer systems are increasingly used by industry and in consumer products because of the environmental pressure to replace solvent-borne polymers. Gels are of particular interest because of their potential use as controlled drug release agents, thickeners and colloidal stabilizers.

In the early eighties we started to work with poly(N-isopropyl-acrylamide) (polyNIPAM) after hearing a presentation by Professor

J. Guillet from the University of Toronto, one of the pioneers of polyNIPAM research[1]. The exciting feature of this polymer is that it is water soluble up to a cloud point temperature (CPT) of 32 °C whereupon it phase separates. Since then many applications of polyNIPAM which exploit its phase separation characteristics have been explored. The list includes photographic film technology[2], colloidal stabilizers[3] and controlled drug release[4]. Most of these applications involve macroscopic gels formed by crosslinking polyNIPAM with methylenebisacrylamide (BA). Our work has concentrated on the preparation and characterization of colloidally dispersed microgel latex based on cross-linked polyNIPAM. The objective of this paper is to summarize our findings.

Figure 1   The structure of polyNIPAM.

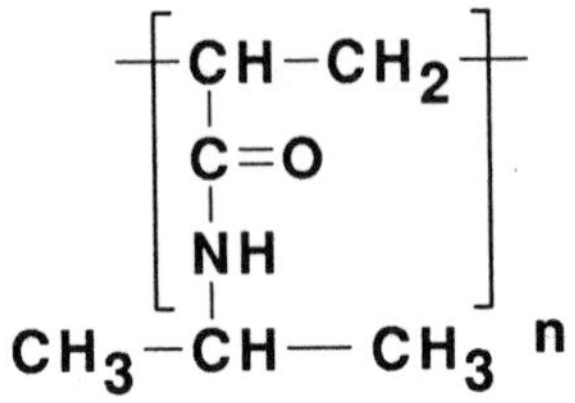

$$\left[\begin{array}{c} CH-CH_2 \\ | \\ C=O \\ | \\ NH \\ | \\ CH_3-CH-CH_3 \end{array}\right]_n$$

PolyNIPAM Homopolymer

Before discussing microgel latex it is relevant to review some key properties of the homopolymer. The water solubility of poly-NIPAM is controlled by the subtle balance between the favorable interactions of water with hydrophilic amide groups and the unfavorable water interactions with hydrophobic propyl groups and backbone methylenes (see Figure 1). Polyacrylamide does not display cloud point behavior whereas polyNIPAM does. Many authors have been employed various techniques to study this phase separation. For example, Figure 2 shows the change in viscosity with temperature as polyNIPAM goes through the cloud point[5]. We offer the following interpretation of this data. In Region II the collapsing chains experienced intermolecular association giving increased viscosity; and in region III the polyNIPAM had completely phase separated, thus contributing very little to the viscosity. Elegant fluorescence studies of polyNIPAM phase separation have been published and are not discussed here[6].

Figure 2   Schematic representation of the viscosity of aqueous solutions of polyNIPAM as a function of temperature[5].

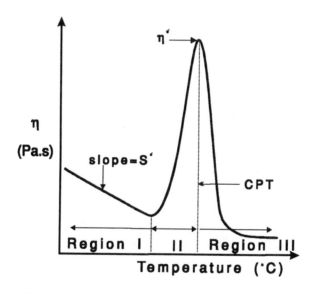

The Interaction of SDS with PolyNIPAM

There are many parallels between the solution properties of polyNIPAM and poly(ethylene oxide), PEO. An important example of this is that polyNIPAM interacts strongly with surfactants by hydrophobic association[7]. In later sections the preparation of latex in the presence of SDS will be discussed so it is relevant to consider the interaction of SDS with polyNIPAM. Figure 3 shows the CPT of 1 wt% polyNIPAM as a function of SDS concentration[8]. The CPT increases with SDS concentration.   Furthermore the CPT of poly-NIPAM in water is rather insensitive to polymer concentration whereas the CPT versus SDS curves such as the one in Figure 3 are very sensitive polymer concentration[9].

Figure 3   Cloud point temperatures for 1 wt% polyNIPAM as functions of SDS concentration.

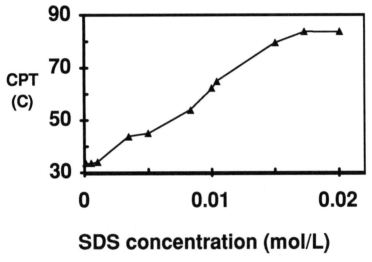

SDS concentration (mol/L)

We have used conductivity and rheology to probe the interaction of SDS with polyNIPAM. Figure 4 shows the first derivative of conductivity as a function of the total concentration of SDS for different polyNIPAM concentrations[9] The curves indicate two apparent transitions; the first, C1, at about 0.001 M SDS, corresponds to the onset of polymer surfactant interaction and the second, C2, has no simple interpretation.   C1 is independent of polymer concentration whereas C2 is proportional to polyNIPAM concentration.

Figure 5 shows that the unusual rheological behavior of polyNIPAM near the CPT (see Figure 2) is attenuated by the presence of SDS[10].   Low concentrations of SDS shifted the peak in the viscosity curve towards higher temperatures whereas the peak disappeared in 0.1 wt% SDS.

Figure 4   Slope of conductivity versus total concentration of SDS measured at 25 °C in presence of 1×10^{-3} M KCl. Wt% polyNIPAM LS12 : ● 0, ▫ 0.05, ◇ 0.15, ◆ 0.2, △ 0.25.

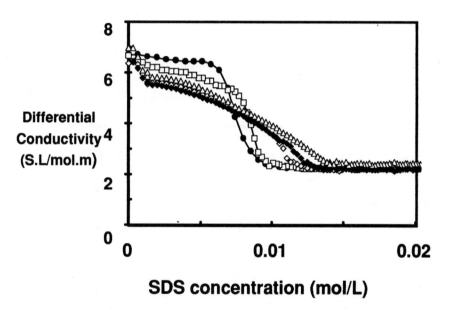

Current evidence suggests that SDS/polyNIPAM aggregates are similar to the model proposed by Cabane et al. for SDS/PEO complexes[11]. That is, spherical micelles form with entrained polymer to give a "string-of-beads" structure.

PolyNIPAM Latex Preparation

   The preparation of polyNIPAM microgel latexes can be understood in terms of two related polymerizations - the aqueous solution polymerization of acrylamide and the surfactant free emulsion polymerization of polystyrene.   Acrylamide is a water soluble monomer which readily undergoes free radical polymerization in water to give high molecular weight water soluble polyacrylamide.   Similarly, NIPAM is also water soluble and gives high molecular weight water soluble polymer if the polymerization is conducted below the CPT of polyNIPAM.

Figure 5   Viscosity of polyNIPAM solutions as functions of temperature for different concentrations of SDS.   Curve labels indicate the SDS concentrations[10].

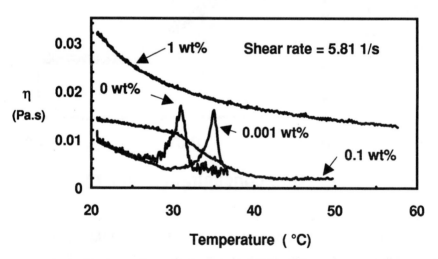

Temperature (°C)

   Introduction of potassium persulfate to a mixture of styrene monomer dispersed in water at 60-70 °C can give monodisperse polystyrene latex with or without surfactant being present[12].   A colloidally stable latex is obtained in the surfactant free case because the sulfate end groups on the polymer chain, originating from the initiator, colloidally stabilize the latex particles. PolyNIPAM microgel latexes have been prepared using the same polymerization conditions but replacing styrene with NIPAM and adding a small amount of methylenebisacrylamide, BA, monomers[13]. The bifunctional BA provides cross-linking and thus prevents the polyNIPAM latex particles from dissolving when the temperature is lowered below the CPT. In this case polyNIPAM is formed above the CPT so that it phase separates to give colloidally stable particles[14].

A transmission electron micrograph of a polyNIPAM latex is shown in Figure 6. The circles are not two dimensional images of spheres but instead are images of the disks of polyNIPAM which remain when the close packed water swollen spheres on the microscope grid are dehydrated in the high vacuum of the electron microscope.

Figure 6   Electron micrograph of polyNIPAM latex. The pattern results from dehydration of a packed array of swollen latex[13].

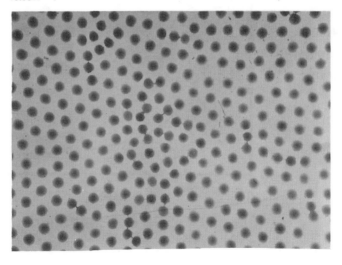

More recently we have shown that the preparations are more robust and average diameters are smaller when the polymerizations are conducted in the presence of low concentrations of SDS. Figure 7 shows the influence of the SDS concentration during polymerization on the particle sizes of the final latexes. It was postulated that the role of the SDS was to increase the colloidal stability of primary particles during the nucleation stage. This effect resulted in a higher concentration of smaller primary particles which ultimately gave a smaller latex. An important restriction for latex formation is that the concentration of SDS must not be too high. Otherwise the CPT of polyNIPAM will be raised above the polymerization temperature and particles will not form.

Figure 7   Average diameter of cleaned polyNIPAM microgel latexes at 25°C as a function of SDS concentration used in the polymerization.

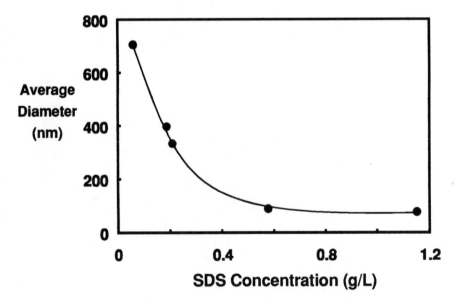

The structure of polyNIPAM latexes is linked to the kinetic of copolymerization of NIPAM and BA monomers. Figure 8 shows the conversion time curve for the monomers at two temperatures. At 70 °C the polymerization was finished after 30 minutes whereas significant monomer remained after 90 minutes of polymerization at 50 °C. At both temperatures the BA was consumed more quickly than the AM. This situation leads to the possibility that the cores of the polyNIPAM particles may have a higher cross-link density than does the exterior of the particles.

Figure 8   Conversion versus time for BA and NIPAM monomers during polyNIPAM latex preparation. Numbers beside labels are polymerization temperature in C.

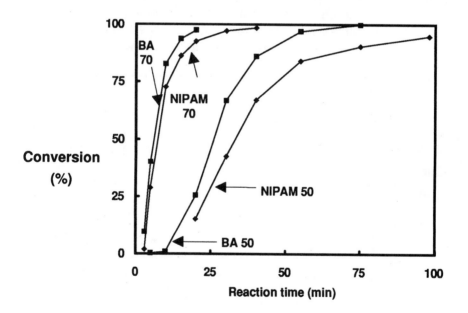

Properties of PolyNIPAM Latex

1. Swelling

The most dramatic property of polyNIPAM latex is that the degree of swelling with water is an extreme function of temperature. Figure 9 shows the diameter of a polyNIPAM latex, measured by dynamic light scattering, as a function of temperature. The average diameter of the polyNIPAM microgel latex decreased by a factor of 2 when the temperature was raised from 10 to 40 °C and the most drastic change occurred at the CPT. Thus, the swelling properties of the crosslinked latex reflects the phase behavior of polyNIPAM homopolymer in aqueous solution.

Figure 9   Average (intensity weighted) diameter of aqueous Latex 05-75 as a function of temperature. The polyNIPAM content of the particles varied with temperature because of temperature dependent swelling with water.

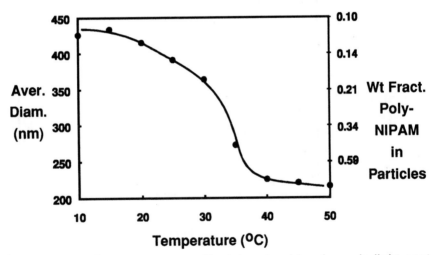

Latex diameters are easily determined by dynamic light scattering. By contrast, the absolute water content of the particles at any temperature is difficult to measure accurately. We have used two methods which yield comparable results. The first was based upon intrinsic viscosity measurements using the Einstein equation to calculate the volume fraction of particles. The second method consisted of centrifuging the dispersion to a close packed bed which was iridescent. The bed was isolated and the water content determined gravimetrically.

Interestingly, under conditions of latex polymerization, 50-70 ^oC, the particles contained more than 10 wt% water. Thus, contrary to recent claims[15], polyNIPAM latex particles can not be considered hydrophobic above the CPT. The presence of water in the poly-NIPAM particles would be expected to promote diffusion of NIPAM monomer and potassium persulfate into particles during the polymerization.

The extent of polyNIPAM microgel latex swelling depends on the BA content of the particles. Figure 10 shows the volume fraction of polyNIPAM in the latex particles at 25 ^oC as a function of the BA content of the latexes. Clearly, the higher the cross-linker concentration, the lower the swelling. The solid line in Figure 10 was calculated from Flory-Huggins theory; see McPhee et al.[14].

Part of the driving force for microgel swelling is the osmotic pressure due to sulfate and carboxyl groups in the polymer network. The charged groups originate from the potassium persulfate initiator used in the polymerization. In our earliest attempts to measure polyNIPAM latex diameter it was apparent that particles were less swollen in the presence of calcium chloride[16]. Figure 11 shows particle size versus temperature at three electrolyte concentrations. Diameters in 0.01 and 0.001 M KCl were similar whereas swelling was less in 0.1 M $CaCl_2$. Note that above the CPT the latexes coagulated in 0.1 M $CaCl_2$ giving anomalous large diameters.

Figure 10   Volume fraction of polymer in swollen latex particles at 25 °C as a function of BA content of gel. The theoretical line was based on Flory Huggins theory[14].

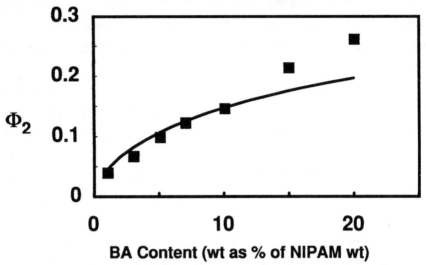

Φ_2

BA Content (wt as % of NIPAM wt)

Figure 11   PolyNIPAM latex diameter as a function of temperature at three electrolyte concentrations. Symbols: ■ 0.1 M CaCl$_2$, ◆ 0.001 M KCl, and ● 0.1 M KCl[16].

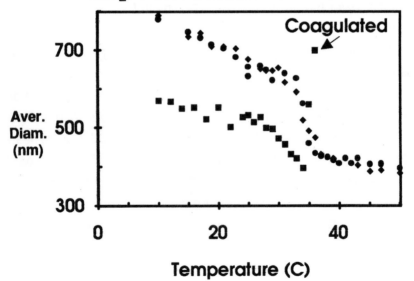

Aver. Diam. (nm)

Temperature (C)

The fundamental driving force for latex swelling is the interaction of the solvent with the polyNIPAM chains. Figures 12 and 13 illustrate how the swelling behavior can be modified by addition of low molecular weight solutes. Addition of methanol gives lower swelling and gives a lower CPT (see Figure 12). By contrast Figure 14 shows that urea also lowers swelling but does not influence the CPT.

Figure 12   Effect of temperature on volume average particle diameter for Latex 05-75L in water and water/methanol mixtures.

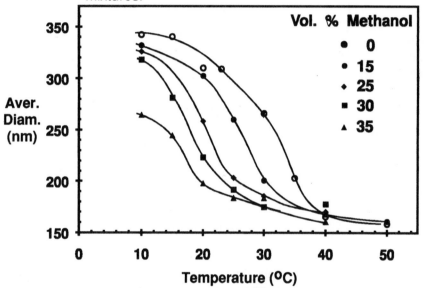

2. Electrical Properties of PolyNIPAM Latex

The charge content of one polyNIPAM latex was measured by conductometric titration and the results are shown in Table 1 together with data from the literature for surfactant free polystyrene latex. Like polystyrene, the polyNIPAM microgel latex has both sulfate and carboxyl groups. However, the charge content of the polyNIPAM is about an order of magnitude lower than that of polystyrene. This is a reflection of the fact that NIPAM polymerizes to very high molecular weight giving few charged end groups compared to polystyrene.

Figure 14   Influence of urea concentration on the swelling/temperature behavior of Latex 05-75L. Diameters were volume weighted averages.

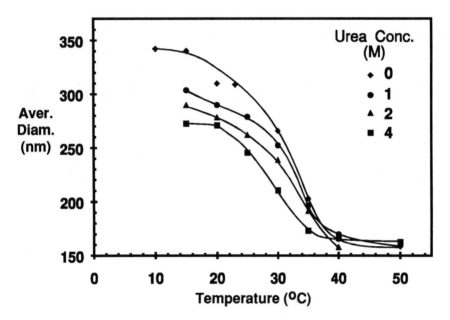

Table 1   Comparison of the charge content of cleaned poly-NIPAM Latex 05-75 with published data for a typical surfactant-free polystyrene latex[14].

Groups	Charge Content (μeq/ g polymer)	
	polyNIPAM Latex	Polystyrene Latex F80 from ref. (12)
sulfate	0.49	40
carboxyl	3.3	363

PolyNIPAM latexes show a spectacular dependency of electrophoretic mobility on temperature in low concentrations of electrolyte. Figure 14 shows the electrophoretic mobility as a function of temperature for two electrolyte concentrations. Below the CPT where the latex was swollen, mobility is very low. Above the CPT mobility becomes more negative to give values more typical of latexes. This is an interesting system because by changing temperature the mobility changes by nearly an order of magnitude whereas the number of electrically charged groups per particle remains constant. Unfortunately, we were unable to determine the locus of the charged groups in the particles. Two extremes are that all the charges were on the surface or all the charges were uniformly distributed throughout the particles.

The temperature sensitivity of the electrophoretic mobility was dampened by in 0.1M $CaCl_2$. Electrolyte effects are shown more clearly in Figure 15. At 40 °C, above the CPT, the electrophoretic mobility shows the expected decrease with increased electrolyte concentration. Indeed, the electrophoretic behavior of this system seems more ideal than does the ubiquitous surfactant free polystyrene latex. Below the CPT at 25 °C there is an anomalous increase in mobility when the ionic strength becomes very low. One explanation of this is that at very low ionic strengths the double layers was sufficiently extended so that charged groups in the interior of the microgel particles can contribute to the mobility.

Figure 14   The electrophoretic mobility of a polyNIPAM latex as a function of temperature. ■ 0.1 M $CaCl_2$, ◆ 0.001 M KCl[16].

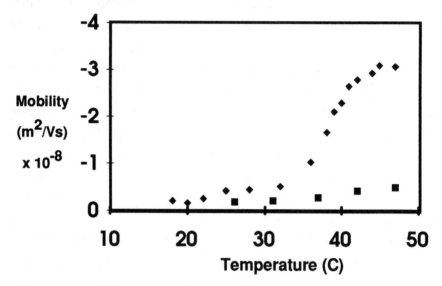

3. Colloid Stability

PolyNIPAM latexes exhibit good colloidal stability. Below the CPT stability comes from steric stabilization, electrostatic stabilization and weak van der Waals attraction. Steric stabilization arises from polyNIPAM tails on the exterior of the particles. Electrostatic stabilization arises from the charged groups originating from the initiator. Finally, the van der Waals forces are expected to be small at low temperatures because the particles contain mainly water. Above the CPT the van der Waals attraction will increase and surface polyNIPAM chains on the particles will be too compressed to give significant steric stabilization. If the electrolyte concentration is high enough to nullify electrostatic stabilization, polyNIPAM latex coagulates when heated past the CPT. An example is shown in Figure 11 where in 0.1 M $CaCl_2$ the latex coagulated as evidenced by increased particle size above 32 °C.

Figure 15   Electrophoretic mobility of Latex 05-75 as a function of
             KCl concentration at two temperatures[14].

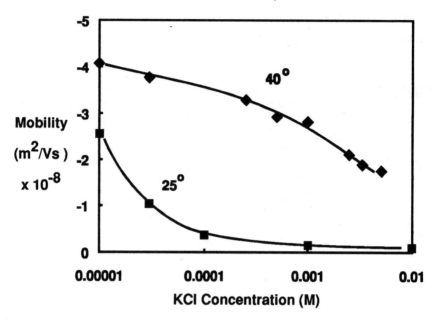

4. Uses

PolyNIPAM microgel latexes offer many exciting potential applications. Our work was initiated to explore the possibility of sterically stabilizing latexes with polyNIPAM to give temperature sensitive stability. We have shown that it is possible to prepare polystyrene latexes with a polyNIPAM shell[3.] An appealing feature of this is that the coagulation temperature can be raised by co-polymerizing acrylamide in the steric shell. Figure 16 shows the critical flocculation temperature of polyNIPAM-co-acrylamide micro-gel latexes as a function of acrylamide concentration used to pre-pare the latex.

strongly above the CPT than below it. They attributed this to enhanced hydrophobic association when the microgel particles dehydrated. Protein which had adsorbed above the CPT, was partially
desorbed upon cooling.

Figure 16   Variation of critical flocculation temperature in 0.1M
             CaCl2 as a function of the amount of acrylamide used
             The corresponding NIPAM charge was 12.5 g/L and
             BA was 1.25 g/L[13].

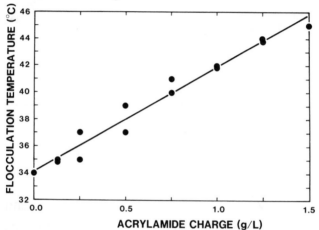

In conclusion polyNIPAM microgel latexes are a unique type
of material offering many potential applications. Remaining scientific
challenges include acquiring a more detailed picture of the kinetics
and mechanism of particle formation together solving with the related question of particle morphology.

Acknowledgments

This work was supported by the Canadian Natural Science
and Engineering Research Council. Support for Michael Tam and
the purchase of some of the equipment was provided by the Mechanical and Chemimechanical Wood-Pulps Network.

1   Heskins, M. and Guillet, J. (1968), J. Macromol. Sci., **A2**(8), 1441.

2   Taylor, L.D. (1960), Novel photographic products and processes, US patent 3,421,892.

3   Pelton, R.H. (1988), J. Polym. Sci., A: Polym. Chem., **26**, 9-18.

4   Hoffman S. (1987), J. Control. Release, **6**, 297.

5   Tam, K.C., Wu, X.W. and Pelton, R.H. (1992), Polym., **33**(2), 436-438.

6   Winnik, F.M., Ringsdorf, H. and Venzmer, J. (1991), Langmuir, **7**, 905 ; Langmuir, **7**, 912.

7   Schild, H. G. and Tirrell, D. A. (1989) Polymer Preprints (ACS Div. Polym. Chem.) **30**(2), 350-51.

8   Wu, X.W. (1992), Ph. D. Thesis, McMaster University, Hamilton.

9   Wu, X.Y., Pelton, R.H., Tam, K.C., Woods, D. R. and Hamielec, A.E. (1992), J. Polym. Sci., Polym. Chem., submitted for publication.

10 Tam, K.C., Wu, X.W. and Pelton, R.H. (1992) J. Polym. Sci., Polym. Chem., submitted for publication.

11 Cabane, B. and Duplessix, R. (1982), J. Physique **43**, 1529.

12 Goodwin, J.W. Hearn, J. Ho, C.C. and Ottewill, R.H. (1974), Br. Polym. J., **5**, 347.

13 Pelton, R.H. and Chibante, P. (1986), Colloids and Surfaces, **20**, 247-256.

14 Wayne McPhee, Kan Chiu Tam and Robert Pelton, J. Colloid Interface Sci., submitted April 1992

15 Kawaguchi, H., Fujimoto, K., and Mizuhara, Y., (1992), Colloid and Polym. Sci., **270**, 53.

16 Pelton, R.H., Pelton, H.M., Morfesis, A. and Rowell, R.L. (1989), Langmuir, **5**, 816-8.

Dielectric Spectroscopic Characterization of a Weak Acid Polystyrene Colloid

by

Lao Sou Su and Sunil Jayasuriya

S.C.Johnson Wax, Racine, WI 53403, USA

Robert M. Fitch*

Fitch & Associates, Racine, WI 53402, USA

Abstract: A commercially available, aqueous polystyrene colloid with surface -COOH groups was purified by exhaustive dialysis. This was then investigated for dielectric frequency response in the low frequency domain (0.01 - 100 Hz). This was done as a function of pH, ionic strength and the chemical nature of the counterions in the electrical double layer. In previous work on strong acid [sulfonate] colloids it had been shown that a plot of the imaginary part of the complex impedance as a function of the real part gave semicircles whose radii were a direct function of the surface charge density. In this work, surface charge is a function of pH. The degree of counterion binding was a function of hydrated ion size, as determined from an analysis of titration data and ionic strength of the continuous phase, which in turn was determined from the phase angle and the characteristic relaxation frequency.

INTRODUCTION

The dielectric spectroscopy of colloidal dispersions in the low-frequency region has been shown to provide information on particle size[1] as well as some of the more complex aspects of the electrical double layer surrounding charged particles[2]. Most previous investigators have found dielectric relaxations in the kilohertz region using colloids of submicron size in solutions of simple electrolyte. Until a few years ago, investigation of lower frequencies was extremely difficult because of interference due to electrode polarization, and because the colloids contained free, inert electrolyte. More recently, highly purified polymer colloids have been prepared which are essentially salt-free[3,4]; and instrumentation is available which overcomes electrode polarization at very low frequencies[5]. As a result, pronounced relaxations readily can be observed at frequencies as low as 10^{-2} Hz, but only when the colloids are totally free of adventitious electrolyte[6]. Typical dielectric loss spectra are shown in Figure 1 for polystyrene colloids carrying sulfonic acid surface groups[6]. Relaxations in this frequency range are presumed to correspond to tangential diffusion of the counterions in the diffuse part of the electrical double layer as an ensemble[7].

EXPERIMENTAL

A model aqueous polystyrene colloid with surface -COOH groups was obtained from the Duke Co. It had a concentration of 0.0526 volume fraction and a nominal

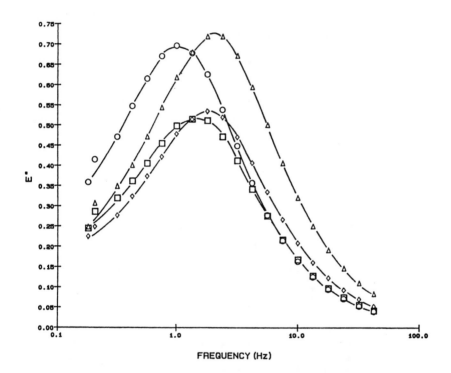

Figure 1: Dielectric permittivity loss as a function of applied frequency. Polystyrene colloids (monodisperse dia. for all is 207 nm), with surface sulfonic acid groups of varying surface charge (C1, 6.6; C2, 12.5; C5, 16.5; C4, 21.1 μC/cm²). Vertical scale in arbitrary units.

diameter of 38 nm. It was purified by exhaustive dialysis and ion-exchange, and was found to have a surface charge density of 11.96 μC/cm². This corresponds to a concentration of RCOOH groups of 0.0103 mol/dm³. The latex was then investigated in the low frequency domain (0.01 - 100 Hz). This was done as a function of pH, ionic strength and the chemical nature of the counterions in the electrical double layer. It was shown previously that for monodisperse colloids with strong acid surface groups a plot of the imaginary part of the complex admittance as a function of the real part gave perfect semicircles, indicating monodispersity, and whose radii were a direct function of the surface charge density. (Figure 2)

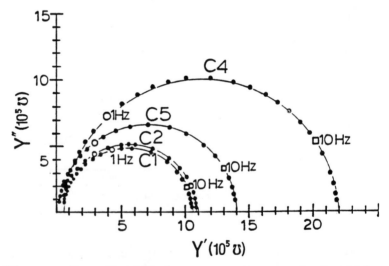

Figure 2: Cole/Cole-type plots of complex admittances of the four polystyrene sulfonic acid colloids shown in Figure 1. Y" and Y' are the imaginary and real parts of the complex admittance, respectively.

Titration

Titration with alkali metal hydroxides leads to simple exchange of H^+ by M^+, where M^+ represents alkali metal cation as shown in Figure 3.

Previous work with strong acid latexes had shown that during titration there is a decrease in the ionic strength of the solution up to the equivalence point. Ionic strength was measured by taking the product of the phase angle, θ, and the logarithm of the characteristic relaxation frequency, ω^*. This phenomenon was attributed to the preferential binding of the alkali metal counterion relative to that of protons. Typical curves for the polystyrene sulfonic acid colloids are shown in Figure 4.

In the present situation, with carboxylated latex, no such decrease was observed, as shown in Figure 5. Our current interpretation of this behavior is that there is no alkali metal ion binding in the case of weak acid

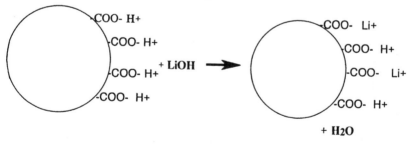

Figure 3: Schematic representation of counterion exchange as a result of titration.

surface groups. The equivalence point is at about pH 8, above which θ log ω^* starts to increase simply because of the increased ionic strength due to the addition of unneutralized M^+OH^-. These differences between strong acid and weak acid latexes are reminiscent of their corresponding conductometric titration behaviors.

The initial value of θ log ω^* indicates, from earlier calibration[6], an ionic strength of 0.013 molar, somewhat higher than that obtained by calculation from the surface charge of the colloid alone. Unlike previous results, this may be due to a contribution from the fixed ions because of the very small size of these latex particles, much smaller than any we have investigated

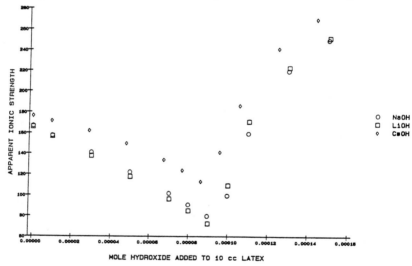

Figure 4: Titration of a polystyrene sulfonic acid latex with three different alkali metal hydroxides. Vertical scale: θ log ω^* = apparent ionic strength, in arbitrary units.

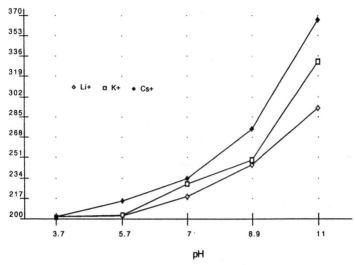

Figure 5: Titration of polystyrene carboxylate latex with three alkali metal hydroxides. Vertical scale: $\theta \log \omega^*$ in arbitrary units.

heretofore. For example, if we model the latex particles as "mega-anions," each would have approximately 3400 charges, and the ionic strength would be about 17 molar. Further experiments with particles of various sizes should resolve this issue.

Conductances

The difference in behavior, seen in Figure 5, of the three alkali metal ions is in the same order as that found in strong acid polystyrene colloids, i.e. Cs^+ has the highest values for $\theta \log \omega^*$. This this can be correlated to its having the highest ionic mobility in water, but more likely, to its being less strongly bound to the surface anions, and thus contributing most to the ionic strength of the solution. The former is supported by the solution conductances we have measured for these systems, as shown in Table I.

TABLE I

Conductance x 10^4 (mho)

pH	3.7	5.7	7.0	11.0
H^+	3.30			
Li^+		3.34	3.94	10.8
K^+		3.38	4.41	19.3
Cs^+		3.64	4.67	27.3

Cesium ion is by far the largest in radius in a crystal lattice, but is the smallest in aqueous solution because it is the least hydrated. This explains its ionic mobility as manifested in its high conductance, especially in this colloid at high pH where an excess of the ions exists in the continuous phase.

Characteristic Relaxations

As is well known from Gouy-Chapman theory, when neutral electrolyte is added to a charged colloid, the diffuse part of the electrical double layer becomes compressed. The low frequency dielectric relaxation, if it refects the response of the diffuse layer as an ensemble, therefore should be drastically affected by the addition of salt. This can be in the form of M^+OH^- beyond the equivalence point in a titration. As the double layer becomes more compact, its characteristic relaxation should be at higher frequencies. Figure 6 shows this is the case.

All of the above experiments have been conducted at extremely low applied potentials of 30 mV/cm. Under these conditions the dielectric permitivity, ε'', of the colloids exhibits a maximum at low frequencies, referred to as the characteristic frequency, and it is this which has been plotted in Figure 6. When the applied field is 300 mV/cm, there is still a maximum in ε'' at about the same frequency, but at 1000 mV/cm there is none. These are shown in Fig 7. Thus we are looking at an extremely delicate ensemble motion of the ions with an energy of a few kT, and which is destroyed at much higher externally applied fields.

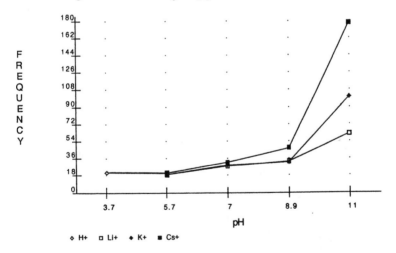

Figure 6: Characteristic relaxation frequency as a function of pH (and thus at high pH, of ionic strength).

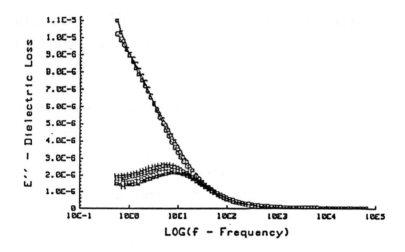

Figure 7: Dielectric loss spectra at different applied potentials: A,a - 30 mV; B,b - 300 mV; C,c - 1000 mV (A,B,C decreasing; a,b,c increasing frequency). Vertical scale in arbitrary units.

CONCLUSIONS

The dielectric behavior of this weak-acid polystyrene colloid shows pronounced differences from that of corresponding strong acid colloids investigated by us earlier, along with several similarities. That the deionized colloids show characteristic relaxations at only a few Hz was first predicted by Schwarz[7]. Most recently Yoshino, in a paper presented at a symposium in Fukui, Japan[8], has treated the ensemble of counterions in the diffuse layer as a spherically symmetrical quantized oscillator. The deformation of a Debye sphere of radius R_0 is given by Yoshino as:

$$R(\theta,\phi,v,t) = R_0(1 + \sum_l\sum_m(\alpha lm(r,t) + \beta lmE)Y^*lm(\theta,\phi)$$

where $Y^*lm(\theta,\phi)$ represents the spherical surface harmonics, $\alpha lm(r,t)$ represents the deformation parameters, βlm is the resonance factor, and E is the external electrical field. On the basis of this theory, Yoshino obtains values of v, the characteristic relaxation frequency for the lowest energy modes, of a few Herz.

Of greatest interest in our results, perhaps, is the fact that one can "see" all of the carboxyl protons, which are undoubtedly solvated by water, in the RCOOH form even though it is a weak polyacid, and therefore only very slightly ionized. Apparently the applied external electrical field can distort the proton cloud sufficiently to obtain a measurable energy loss within its characteristic relaxation frequency domain. A comparison of Figs. 1 and 7 shows that the characteristic frequency of the RCOOH colloid is about 10 times those of the RSO_3H latexes. This may be due to the smaller particle size (smaller R_0). Also, there is apparently no alkali metal counterion binding in the $RCOO^-M^+$ type of latex, in contrast to that in the $RSO_3^-M^+$ type.

Dielectric spectroscopy thus offers a non-destructive means for investigating very low energy phenomena in the electrical double layers of colloids.

REFERENCES

1. Schwan, H.P., Schwarz, G., Maczuk, J. and Pauly, H. (1962). J. Phys. Chem., **66**, 2626 .

2. Rosen, L.A. and Saville, D.A. (1990). J. Colloid Interface Sci. **140** (1), 82 .

3. Chonde, Y. and Krieger, I.M. (1980). J. Colloid Interface Sci. **77**, 138.

4. Tsaur, S.L. and Fitch, R.M. (1987). J. Colloid Interface Sci. **115**, 450

5. Solartron   Instruments, Schlumberger Corp., Victoria Rd., Farnborough, Hampshire, England.

6. Fitch, R.M., Su, L.S. and Tsaur, S.L. (1990). In Scientific Methods for the Study of Polymer Colloids and their Applications   (Eds. F. Candau, and R.H. Ottewill) , pp. 373 - 391, The Netherlands, Kluwer Academic Publishers.

7. Schwarz, G., (1962). J. Phys. Chem. **66**, 2636.

8. Yoshino, S. (1993). Proceedings of the International Symposium on Polymeric Microspheres, Fukui, Japan, Paper No. N-49, in Polymer International, in press.

INTERFACE THERMODYNAMICS IN
COMPOSITE LATEX PARTICLES

J A Waters

ICI Paints,
Slough, UK.

INTRODUCTION

Composite latex particles with a wide diversity of internal
morphology have been described in the literature. There
have been many papers and patents and a number of reviews
(Eliseeva, 1985; Daniel, 1985). Particle structures have
included one polymer with multiple internal inclusions of
another polymer; "core-shell" particles with one polymer
forming a more or less complete layer around the other and
various non-spherical composites with two different
polymers partially interfacing with the continuous phase.

Several factors which are encountered during preparation
of the composite particles, are believed to influence the
particle morphology. Some workers have pointed to the
importance of interfacial energies (Berg et al, 1986, 1989;
Dimonie et al, 1989; Sundberg et al, 1990). This theme is
explored in this paper. The derivation will be given for
relationships between the interfacial energies and the
relative volumes of the two dissimilar polymers, and the
use of these relationships to predict favoured particle

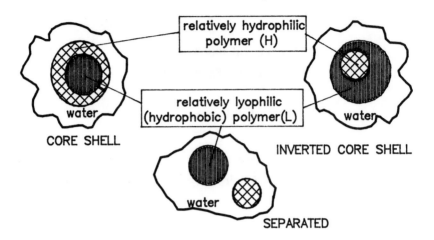

Fig. 1.   The three extreme particle morphologies

morphologies will be shown. An extension of this
theoretical treatment leads to equations which can be used
to identify favoured intermediate structures where one
polymer is partially enveloping or engulfing the other.

EXTREME PARTICLE MORPHOLOGIES

In aqueous systems, three extreme particle morphologies
may be defined (Waters, 1989, 1991). CORE-SHELL has the
relatively more hydrophilic polymer (H) forming a complete
layer around a particle of the other polymer; INVERTED
CORE-SHELL is the inverse arrangement with the relatively
more hydrophobic or lyophilic polymer (L) constituting the
shell and interfacing with the aqueous continuous phase;
and SEPARATED where the two polymers exist as separate
particles (Fig. 1).

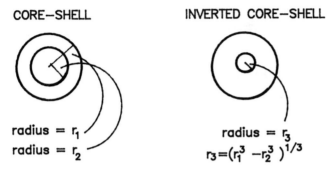

CORE–SHELL   INVERTED CORE–SHELL

radius = r_1
radius = r_2

radius = r_3
$r_3 = (r_1^3 - r_2^3)^{1/3}$

Fig. 2. Inter-relationship of the interfacial areas, $A_i = 4\pi r_i^2$

Calculation of Interfacial Energy

For each of these morphologies, the total interfacial energy is given by (Berg et al, 1986).

$$E = \Sigma \gamma_i A_i \qquad (1)$$

where A is the interfacial area and γ is the interfacial energy. For the CORE-SHELL (Fig. 2), this can be written as

$$E_{cs} = 4\pi [r_1^2 \gamma_{H-L} + r_2^2 \gamma_{H-w}]$$
$$= 4\pi [(3v_L/4\pi)^{2/3} \gamma_{H-L} + (3V/4\pi)^{2/3} \gamma_{H-w}]$$

where v_H, v_L are the volumes of the relatively hydrophilic and hydrophobic (lyophilic) polymers respectively and

$$V = v_H + v_L$$

Then
$$E_{cs} = 3^{2/3}(4\pi)^{1/3}[v_L^{2/3} \gamma_{H-L} + V^{2/3} \gamma_{H-w}] \qquad (2)$$

For the INVERTED CORE-SHELL

$$E_{INV} = 3^{2/3} (4\pi)^{1/3} [v_H^{2/3} \gamma_{H-L} + V^{2/3} \gamma_{L-W}] \qquad (3)$$

and for the SEPARATED morphology

$$E_{SEP} = 3^{2/3} (4\pi)^{1/3} [v_L^{2/3} \gamma_{L-W} + v_H^{2/3} \gamma_{H-W}] \qquad (4)$$

Derivation of Relationships to Predict Preferred Extreme Morphology

The CORE-SHELL has lower interfacial energy and is favoured over the INVERTED CORE-SHELL if

$$E_{CS} < E_{INV}$$

$$v_L^{2/3} \gamma_{H-L} + V^{2/3} \gamma_{H-W} < v_H^{2/3} \gamma_{H-L} + V^{2/3} \gamma_{L-W}$$

$$\gamma_{L-W} V^{2/3} - \gamma_{H-W} V^{2/3} > \gamma_{H-L} (v_L^{2/3} - v_H^{2/3})$$

$$(\gamma_{L-W} - \gamma_{H-W})/\gamma_{H-L} > (v_L^{2/3} - v_H^{2/3})/V^{2/3}$$

It is convenient to define a fractional volume (ν) such that

$$\nu_H = v_H/V ; \quad \nu_L = v_L/V$$

and

$$\nu_H + \nu_L = 1$$

That is, the total volume is considered to be unity. The condition under which CORE-SHELL is preferred over INVERTED CORE-SHELL is given by

$$(\gamma_{L-W} - \gamma_{H-W})/\gamma_{H-L} > \nu_L^{2/3} - \nu_H^{2/3} \qquad (5)$$

Similarly, it can be shown that if

$$E_{CS} < E_{SEP}$$

then

$$v_L^{2/3}\gamma_{H-L} + v^{2/3}\gamma_{H-w} < v_L^{2/3}\gamma_{L-w} + v_H^{2/3}\gamma_{H-w}$$

substituting v values

$$V^{2/3}v_L^{2/3}\gamma_{H-L} + V^{2/3}\gamma_{H-w} < V^{2/3}v_L^{2/3}\gamma_{L-w} + V^{2/3}v_H^{2/3}\gamma_{H-w}$$

$$\gamma_{L-w}v_L^{2/3} - \gamma_{H-L}v_L^{2/3} > \gamma_{H-w}(1- v_H^{2/3})$$

and CORE-SHELL is preferred over SEPARATED if

$$(\gamma_{L-w} - \gamma_{H-L})/\gamma_{H-w} > (1- v_H^{2/3})/v_L^{2/3} \tag{6}$$

Similarly INVERTED CORE-SHELL is preferred over SEPARATED if

$$(\gamma_{H-w} - \gamma_{H-L})/\gamma_{L-w}) > (1- v_L^{2/3})/v_H^{2/3} \tag{7}$$

The relationships (5) to (7) (Waters, 1989, 1991) have expressions of the fractional volumes on the right-hand-side. Values for the expressions range from -1 to +1 for relationship (5) (Fig. 3) and from 0 to +1 for the relationships where SEPARATED is an option (Fig. 4).

Morphology Prediction

The extreme morphology with the lowest interfacial energy is identified by considering two of the relationships.

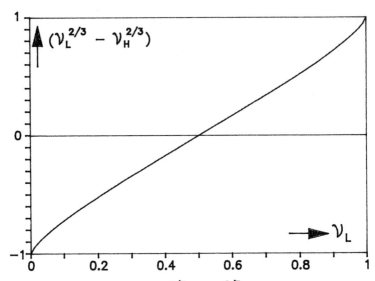

Fig. 3.   *The curve for* $(\mathcal{V}_L^{2/3} - \mathcal{V}_H^{2/3})$

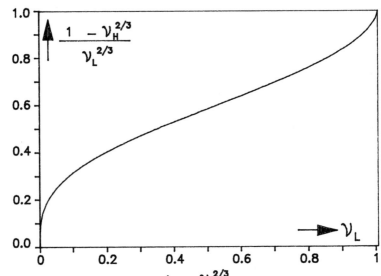

Fig. 4.   *The curve for* $\dfrac{1 - \mathcal{V}_H^{2/3}}{\mathcal{V}_L^{2/3}}$

Systems where $\nu_L < 0.5$. From relationship (5), it can be seen that if $\nu_H > \nu_L$, the RHS expression is negative, and because by definition $\gamma_{L-w} > \gamma_{H-w}$, the LHS expression is positive and CORE-SHELL is preferred over INVERTED CORE-SHELL as expected intuitively.

To test whether CORE-SHELL is preferred over SEPARATED, relationship (6) is considered. The relationship may or may not be fulfilled depending on the relative values of the three interfacial energies. There is a high probability that CORE-SHELL is preferred at low ν_L because the RHS has low value (Fig. 4) and the LHS would be expected to have a value of approximately 1 at the least.

System A. If a polymer H is relatively polar, such as poly (methyl methacrylate) and polymer L is relatively hydrophobic, such as poly(butyl acrylate), the values for the interfacial energies at 60°C might be as assigned (Table 1). The interfacial energy expression in relationship (5) has a value greater than 1 indicating that CORE-SHELL is preferred over INVERTED CORE-SHELL for all values of ν_L. Similarly, with the interfacial energy expression in relationship (6) having a value in excess of 1, CORE-SHELL is preferred over SEPARATED for all compositions.

If the preparation of composite latex particle preparation was attempted by firstly preparing particles of the polymer L and then adding monomer and polymerising to produce the polymer H (polymer 2), the value for ν_H would be very small in the early stages, and even at these low values, a CORE-SHELL structure with a very thin shell of polymer H would be preferred.

TABLE 1.

Interfacial energies for systems A, B, C

system	γ_{L-w}	γ_{H-w}	γ_{H-L}	X	Y	Z
A	33	17	3.5	1.7	4.6	0.41
B	17	7	3.5	1.9	2.9	0.21
C	7	5.5	3.5	0.64	0.43	0.29

$$X = (\gamma_{L-w} - \gamma_{H-L})/ \gamma_{H-w}$$
$$Y = (\gamma_{L-w} - \gamma_{H-w})/ \gamma_{H-L}$$
$$Z = (\gamma_{H-w} - \gamma_{H-L})/ \gamma_{L-w}$$

System B. Emulsion polymerisation to produce colloidally-
stable polymer particles is often carried out using a
surfactant which includes poly(ethylene oxide) (PEG) to
provide steric stabilisation. The solvated layer of PEG
at the particle surface would be expected to give a
substantial reduction in the interfacial energy. Particle
preparation may be considered as in system A, but starting
with particles (polymer 1) of the intrinsically more hydrophobic polymer having a surface layer of PEG or other bound
hydrophilic material to render them more hydrophilic to an
extent where the second polymer becomes (L). If, in such a
system, the appropriate interfacial energy expressions have
values greater than 1 (Table 1), the preferred structure
comprises a domain of polymer 2 within the polymer 1
particle (CORE-SHELL), and this arrangement would remain
favoured even if (v_L) approached 1.

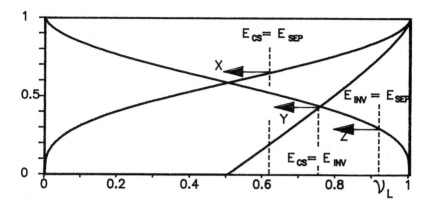

Fig. 5. System (C) : compositions where two extreme morphologies have the same interfacial energy

System C. If, however, the composite particle preparation as in System B was repeated, but under conditions where on contact with the aqueous phase, polymer 2 was coated with PEG or other hydrophilic material (to become Polymer H), the two appropriate interfacial energy expressions could have values less than 1.

With the interfacial energies as assigned (Table 1) and using the relationships (5) to (7), it can be concluded that of the extreme morphologies, CORE-SHELL has the lowest interfacial energy where $v_L < 0.6$, that INVERTED CORE-SHELL has the lowest interfacial energy when $v_L > 0.92$ and that SEPARATED is preferred for all other values of v_L (Fig. 5). However it might be expected that an intermediate structure with both polymers partially interfacing with the continuous phase, would have the lowest total interfacial energy. This is confirmed below for system C.

Evaluating the Interfacial Energy Expression

To test the relationships (5) to (7), it is not essential
to have values for the three individual interfacial
energies. A value for the entire interfacial expression
may be obtained by constructing a fully representative
model and measuring the contact angle (Waters, 1989,1991).
The method involves identifying a solvent for one of the
polymers (B) which has the same surface energy as the
polymer, and preparing a solution. Surfactants which are
associated with the polymer in the real system, must be
included. A plaque or film of the other polymer (A) is
constructed and placed under water, pre-saturated with the
solvent. The surface of polymer A is conditioned by
addition of surfactant or other components which are
associated with the surface of polymer A in the real
system. The solution of polymer (B) is used to form a
droplet on the surface of polymer (A) (Fig. 6). From the
Young-Dupre equation (Young, 1804)

$$(\gamma_{A-W} - \gamma_{A-B})/\gamma_{B-W} \quad = \quad Cos \ \theta$$

for $0 < \theta < \pi$. If complete spreading is observed ($\theta = 0$),
the value for the interfacial energy expression is at least
1 and because the maximum value for the fractional volume
expression is 1, it is confirmed that the relationship is
fulfilled.

INTERMEDIATE MORPHOLOGIES

Doublet particles with both polymers interfacing with the
diluent phase, may be considered to have a morphology

intermediate between the CORE-SHELL (or INVERTED CORE-SHELL) and SEPARATED morphologies. A geometrical analysis of the doublets is given below, where one polymer (Q) is considered to be a non-deformable sphere and the other polymer (P) is able to envelop or engulf the first, and to minimise its interfacial area against water. This analysis can be used to identify which intermediate structure or extreme structure has the lowest total interfacial energy. Also it predicts how composite particle morphology will rearrange if the particles are made under non-equilibrium conditions and the conditions are subsequently altered to permit rearrangement. An analysis of a doublet system, comprising immiscible oils, has been made by considering spreading coefficients (Torza and Mason, 1970). Polymer doublets have been considered (Chen et al, 1992).

The following analysis is based on

$$E = \Sigma A_i \gamma_i \qquad (1)$$

and requires determination of the interfacial areas (A).

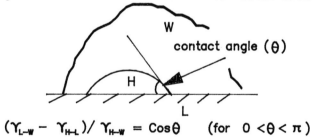

$$(\gamma_{L-w} - \gamma_{H-L}) / \gamma_{H-w} = \cos\theta \quad \text{(for } 0 < \theta < \pi\text{)}$$

Fig. 6. Measurement of the interfacial energy expression

Derivation of General Equation

A composite particle having a total volume of unity is
considered. If the non-deformed spherical core polymer (Q)
of volume v_Q, is fully enveloped or engulfed by the other
polymer (P) of volume v_p, then the total overall radius is
given by

$$r = (3/4\pi)^{1/3}$$

the radius of the core

$$r_Q = (3v_Q/4\pi)^{1/3}$$

the interfacial area against the continuous phase

$$A_{P-W} = 4\pi(3/4\pi)^{2/3} = 2(\pi/2)^{1/3}.3^{2/3} \qquad (8)$$

and the interfacial area between the two polymers

$$A_{P-Q} = 2(\pi/2)^{1/3}(3v_Q)^{2/3} \qquad (9)$$

From equations (1), (8), (9), the total interfacial
energy for the fully engulfed structure (which can be
either CORE-SHELL or INVERTED CORE-SHELL as defined above)
is given by

$$E_o = 2(\pi/2)^{1/3} 3^{2/3} [\gamma_{P-W} + v_Q^{2/3}\gamma_{P-Q}] \qquad (10)$$

If the composite particle is a doublet where the core is
not fully engulfed by the other polymer, the interfacial
areas can be expressed in terms of the angle (ϕ_Q) subtended
by the line joining the particle centres and the line from
the centre of the core particle to a point where the three
phases meet (Fig.7). The three interfacial areas are the
curved surfaces of truncated spheres. Expressions for
these areas are known (Fig.8).

The curved area subtended by the angle (2θ) is given by:

$$A_I = 2\pi r^2 [1-Cos\theta] \qquad (11)$$

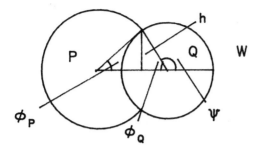

Fig. 7.   Cross-section of doublet.

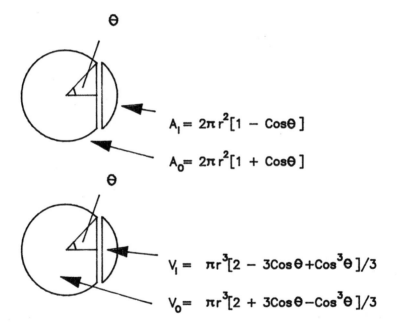

$$A_I = 2\pi r^2[1 - \cos\Theta]$$

$$A_O = 2\pi r^2[1 + \cos\Theta]$$

$$V_I = \pi r^3[2 - 3\cos\Theta + \cos^3\Theta]/3$$

$$V_O = \pi r^3[2 + 3\cos\Theta - \cos^3\Theta]/3$$

Fig. 8.   Curved-surface-areas and volumes for a truncated sphere

and for the other curved surface,

$$A_o = 2\pi r^2 [1+Cos\theta] \tag{12}$$

$$\therefore \quad A_{Q-W} = 2\pi r_Q^2 [1+Cos\phi_Q] \quad \text{where } r_Q = (3v_Q/4\pi)^{1/3}$$

$$A_{Q-W} = (\pi/2)^{1/3} (3v_Q)^{2/3} [1+Cos\phi_Q] \tag{13}$$

and $$A_{P-Q} = (\pi/2)^{1/3} (3v_Q)^{2/3} [1-Cos\phi_Q] \tag{14}$$

The interfacial area (A_{P-W}) is the curved surface area of a truncated sphere which has a volume V_T after truncation, where V_T comprises the sum of the volume of the engulfing polymer (v_P) and the segmental volume of the core polymer which has been engulfed (V_{seg})

$$V_T = v_P + V_{seg} \tag{15}$$

Using equation (12), this truncated sphere has a curved surface area

$$A_{P-W} = 2\pi r_P^2 [1+Cos\phi_P] \quad \text{where } r_P = h/Sin\phi_P$$

$$h = r_Q Sin\phi_Q = (3v_Q/4\pi)^{1/3} Sin\phi_Q \quad \text{(Fig. 8)}$$

$$\therefore A_{P-W} = (\pi/2)^{1/3} (3v_Q)^{2/3} [1+Cos\phi_P] Sin^2\phi_Q / Sin^2\phi_P \tag{16}$$

Determination of ϕ_P. The relationship between ϕ_P and ϕ_Q may be established by considering the volume V_T. Expressions for the volume of truncated spheres can be derived giving

$$V_o = \pi r^3 [2+3Cos\theta-Cos^3\theta]/3 \tag{17}$$

and $\qquad V_I = \pi r^3 [2-3\text{Cos}\theta + \text{Cos}^3\theta]/3 \qquad\qquad (18)$

where V_I is the volume of the truncated sphere subtended by the angle 2θ and V_o is the volume of the remainder of the total sphere (subtended by the angle $2(\pi-\theta)$) (Fig. 8).

From equation (17),

$$V_T = \pi r_p^3 [2+3\text{Cos}\phi_p - \text{Cos}^3\phi_p]/3$$

$$= \pi h^3 [2+3\text{Cos}\phi_p - \text{Cos}^3\phi_p]/3\text{Sin}^3\phi_p$$

It is convenient here to define a function $F(\theta)$ where

$$F(\theta) = (2+3\text{Cos}\theta - \text{Cos}^3\theta)/\text{Sin}^3\theta \qquad\qquad (19)$$

Then $\qquad V_T = \pi h^3 F(\phi_p)/3$

and $\qquad V_T = \nu_p + V_{seg} \qquad\qquad (15)$

From equation (17)

$$V_{seg} = \pi h^3 [2+3\text{Cos}\psi - \text{Cos}^3\psi]/3\text{Sin}^3\psi$$

where $\qquad\qquad \psi = \pi - \phi_Q \qquad\qquad$ (Fig. 7)

$$V_{seg} = \pi h^3 F(\psi)/3$$

From equation (15)

$$\pi h^3 F(\phi_p)/3 = \pi h^3 F(\psi)/3 + \nu_p$$

$$F(\phi_p) = F(\psi) + 3\nu_p/\pi h^3 \; ; \quad h^3 = 3\nu_Q \text{Sin}^3\psi/4\pi$$

$$F(\phi_p) = F(\psi) + 4\nu_p/\text{Sin}^3(\psi)\nu_Q \qquad\qquad (20)$$

After calculating $F(\phi_P)$ according to equation (20), it is possible to determine (ϕ_P) by a process of reiteration or by constructing a table of values for $F(\phi_P)$ for values of (ϕ_P), (using equation 19).

The Total Interfacial Energy. For any given value of ν_Q (or ν_P) and given values of the three interfacial energies, the total interfacial energy may be calculated for all values of ϕ_Q. After determining ϕ_P (above), and using equations (1), (13), (14) and (16), the total interfacial energy is given by:

$$E = (\pi/2)^{1/3}(3\nu_Q)^{2/3}N$$

where N is given by

$$\gamma_{P-W}(1+Cos\phi_P)Sin^2\phi_Q/Sin^2\phi_P + \gamma_{Q-W}(1+Cos\phi_Q) + \gamma_{P-Q}(1-Cos\phi_Q)$$
$$(21)$$

and from equation (10)

$$E/E_o = \nu_Q^{2/3}N/[2(\gamma_{P-W} + \nu_Q^{2/3}\gamma_{P-Q})]$$

$$E/E_o = N/[2(\gamma_{P-Q} + \gamma_{P-W}/\nu_Q^{2/3})] \qquad (22)$$

where E_o is the interfacial energy for the fully ENGULFED structure ($\phi_Q = \pi$).

Plots of E/E_o versus ϕ_Q indicate whether an intermediate doublet morphology has lower total interfacial energy than both the ENGULFED STRUCTURE (either CORE-SHELL or INVERTED CORE-SHELL) and the SEPARATED structure ($\phi_Q = 0$). The degree of engulfment corresponding to the lowest energy can be identified. The steepness of the curves indicates the magnitude of the thermodynamic driving force for rearrangement of the morphology.

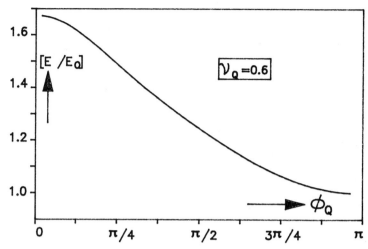

Fig. 9. *Variation of interfacial energy with*
engulfment : system (B) with polymer (L) as Q

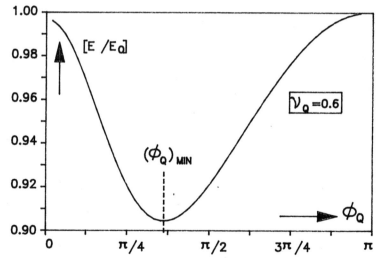

Fig. 10. *Variation of interfacial energy with*
engulfment : system (C) with polymer (L) as Q

Example Systems

System (B) (Table 1) may be considered with polymer (L) as
the non-deforming sphere (Q) and v_Q = 0.6. The curve for
E/E_o versus ϕ_Q does not show a minimum (Fig. 9). A steady
decrease in interfacial energy with increasing ϕ_Q, that is
with progressive engulfment, is indicated. Also it can be
seen that the system experiences a 40% reduction in
interfacial energy when going from SEPARATED to ENGULFED
(CORE-SHELL).

By contrast, system (C) (Table 1) with polymer (L) as
(Q) and v_Q = 0.6 shows a minimum at ϕ_Q = 0.36π (65°) (Figs.
10,11) and this intermediate doublet (Fig. 11) is indicated
as the preferred structure. For this system at v_Q = 0.6,
the interfacial energies for SEPARATED and ENGULFED are the
same (expected because LHS = RHS in relationship (6)). If
system (C) is considered but with polymer (H) as (Q) and v_Q
= 0.4, the plot indicates a minimum at ϕ_Q = 0.34π (61°) and
confirms that the ENGULFED structure (now INVERTED CORE-
SHELL) has higher interfacial energy (and is less favoured)
compared to SEPARATED (Fig. 12)).

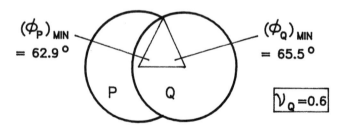

Fig. 11. System (C) at minimum interfacial energy
with polymer (L) as Q

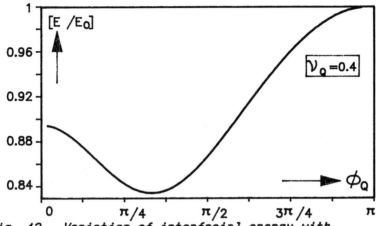

*Fig. 12.   Variation of interfacial energy with
engulfment : system (C) with polymer (H) as Q*

With the interfacial energies as in system (C), the
growth of a non-deforming domain may be considered by
plotting curves from low ν_Q to high ν_Q. This has been done
with polymer(L) as (Q) (Fig.13). The plots indicate that
with growth of the domain a larger proportion of the domain
surface interfaces with the water continuous phase. That
is, with growth, the hydrophobic domain will appear to be
emerging from the enveloping polymer (Fig.14). It can be
shown that, if relatively hydrophilic, the domain will
emerge more rapidly.

A similar predictive method can be used with composite
particles having multiple domains. One domain is
considered; the other domains serve to increase the
effective volume of the engulfing polymer particle and
thereby decrease the operative value for ν_Q (Fig.15). This
increases the likelihood that ENGULFED will be preferred
over SEPARATED, irrespective of whether the former is
CORE-SHELL or INVERTED CORE-SHELL.

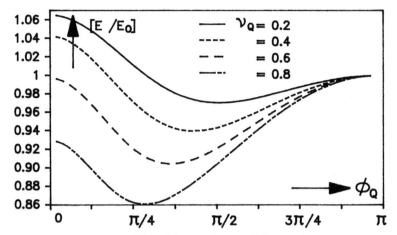

Fig. 13. System (C) with polymer (L) as Q.
Variation of $\mathcal{V}_Q$

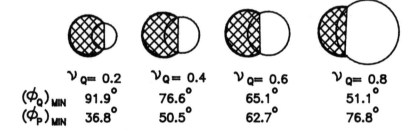

$\mathcal{V}_Q = 0.2$	$\mathcal{V}_Q = 0.4$	$\mathcal{V}_Q = 0.6$	$\mathcal{V}_Q = 0.8$
$(\phi_Q)_{MIN}$ 91.9°	76.6°	65.1°	51.1°
$(\phi_P)_{MIN}$ 36.8°	50.5°	62.7°	76.8°

Fig. 14. Diagram depicting emergence of domain (Q) with growth

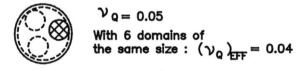

$\mathcal{V}_Q = 0.05$

With 6 domains of
the same size : $(\mathcal{V}_Q)_{EFF} = 0.04$

Fig. 15. Multiple domains effectively decrease $\mathcal{V}_Q$

Rewriting the relationships (6) and (7) in terms of P and Q, the condition under which ENGULFED is preferred over SEPARATED is

$$(\gamma_{Q-W} - \gamma_{P-Q})/\gamma_{P-W} > (1- \nu_P^{2/3})/\nu_Q^{2/3} \qquad (22)$$

When the value for the interfacial energy expression increases, the curves of E/E_o against ϕ_Q are less likely to show a minimum and they give a steeper decline to ENGULFED. Such an increase arises when γ_{Q-W} increases (Q becomes more hydrophobic); γ_{P-Q} decreases; or γ_{P-W} decreases (P becomes more hydrophilic). This is illustrated for systems similar to system (C), but where γ_{P-W} is decreased (Fig.16). The curves show that the minimum is not observed when the value for the interfacial energy expression is (≥ 1); that is, at these values, ENGULFED is preferred over SEPARATED and all INTERMEDIATE structures.

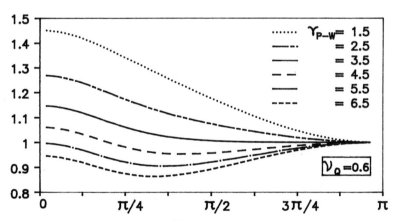

Fig. 16. Variation of γ_{P-W} with $\gamma_{Q-W} = 7$; $\gamma_{P-Q} = 3.5$

SUMMARY AND CONCLUSIONS

This theoretical development has concerned equilibrium structures and is based on the minimisation of interfacial energy.

For the three extreme morphologies, relationships have been derived between the three interfacial energies and the relative volumes of the two polymers. The preferred extreme morphology can be identified by assessing two of the relationships.

An equation has been derived which shows that for some composite particle systems, an intermediate structure with both polymers partially interfacing the continuous phase will be preferred. By altering one or more of the inter-facial energies, it is possible to favour engulfment of one polymer by the other and to increase the thermodynamic driving force for engulfment. The equation predicts the changes in the preferred intermediate structure when the volume of one component increases.

ACKNOWLEDGEMENTS

The author expresses gratitude to Professor R H Ottewill for numerous stimulating discussions and encouragement in this and other areas of polymer colloids, and he thanks Mr J S Keith whose astute observations with composite latex particles prompted this theoretical analysis. The author is grateful to ICI Paints, for permission to publish this paper.

REFERENCES

Berg, J., Sundberg, D., Kronberg, B., (1986). Polym. Mat.
Sci. Eng., ACS Symp. Series, 54.

Berg, J., Sundberg, D., Kronberg, B., (1989). Microencap-
sulation, 6, 327.

Chen, Y.C., Dimonie, V., El-Aasser, M.S., (1992). J.
Applied Polymer Science, 45, 487.

Daniel, J.C., (1985). Makromol. Chem. Suppl., 10/11, 359.

Dimonie, V.L., Chen, Y.C., El Aasser, M.S., Vanderhoff,
J.W., (1989). "Deuxieme Colloque International sur les
Copolymerisations et les Copolymeres en Milieu
Disperse", Lyon, 81.

Eliseeva, V.I., (1985). Prog. Organic Coatings, 13, 195.

Lee, S., Rudin, A., (1992). Advances in Chemistry, ACS
Symp. Series 492, Ch.15, 234.

Sundberg, D., Casassa, A.P., Pantazopoulos, J., Muscato,
M.R., Kronberg, B., Berg, J., (1990). J. Applied Polymer
Science, 41, 1425.

Torza, S., Mason, S.G., (1970). J. Colloid Interface Sci.
33, 67.

Waters, J.A., to ICI plc, (1989). European patent
application 0327,199 A2.

Waters, J.A., to ICI plc, (1991). US Patent 4,997,864.

Young, T., (1804). Proc. Roy. Soc. (London).

Permeation through Polymer Latex Films made from Model Colloids.

J. Hearn, P.A. Steward, M. Chainey[1], B.J. Roulstone[2],
Dept. of Chemistry & Physics, Nottingham Polytechnic, Clifton
Lane, Nottingham NG11 8NS.

M.C. Wilkinson.
C.B.D.E. Porton Down, Salisbury, Wiltshire SP4 0JQ.

[1] Currently at Courtaulds Coatings (Holdings) Ltd., Gateshead.
[2]Currently at I.C.I. Paints, Slough.

Introduction.

On the grounds of cost, safe handling, and environmental acceptability, aqueous latex coatings are to be preferred to organic solvent based systems. These advantages outweigh the problems of greater energy requirements, arising from the higher latent heat of vaporisation of water, and poorer film properties such as lower gloss and greater permeability to pollutant gases. In spraying applications latex can be the preferred choice, even when water is a potential solvent, because of the low viscosity and high solids loading of the latex, (pseudo latex), formulation [1]. The aim of this paper is to review briefly the background theories of film formation from polymer latices and to review results obtained for transport through films made from "model colloid" latices in relation to their morphology.

Film formation.

Upon evaporation of water, latex particles concentrate and may dry to form either a transparent tough continuous film or a friable opaque powder-compact depending upon the temperature relative to the minimum film formation temperature, (MFFT). In fact the former desirable outcome is something of a compromise since the tendency of the polymer spheres to flow and fuse into a continuous film can at the other extreme result in a permanently tacky film more suited to adhesive applications [2].

Whilst film formation from latices and from polymers in solution seem to be fundamentally different processes there are aspects of similarity, eg macromolecules in solution behave hydro-dynamically as though they are molecular dispersions having solvent impermeable cores and peripheral solvent permeable segments [3], a difference mainly of scale compared with uncharged sterically stabilized latex particles. Whether solvents will necessarily deposit pore-free films of the maximum density and lowest permeability is uncertain. Different outcomes are predicted, depending upon the solvent power, when high concentrations are reached in the later stages of drying. Funke [4] suggests, based upon evidence from freeze-dried extracts of films in the gel stage, that good solvents would produce more open porous structures and Nicholson [5] sites Gould [6] with a similar prediction that compact molecules in the solution will remain compact in the film state. Kesting [7], however, suggests that polymer coils in good solvents will interpenetrate to a more compact structure than in poorer solvents which will favour earlier polymer segment-segment contacts.

The rate of evaporation in solvent casting depends upon time $t^{1/2}$ as expected for a process controlled or limited by diffusion, ie. Fickian diffusion to the surface through a homogeneous solution of increasing concentration [8]. The removal of the final traces of solvent from these films is a problem since they may be plasticised by the solvent. Elevated temperatures, good vacuum and long times are used to overcome the problem, eg. 50° C, 3 Torr, 96 hrs [9].

Several studies [10, 11, 12, 13] of water evaporation rates in latex film formation suggest that three stages are involved:

i.   Firstly, water evaporates at a constant rate virtually unaffected by the presence of the particles.

ii.   In the second stage, the rate of evaporation falls rapidly when the particles come into irreversible contact and evaporation occurs from a reducing air-water interfacial area during particle deformation and capillary channel closure.

iii.   Finally, the third stage shows very slow weight loss ascribed to water diffusion through the polymer matrix.

Croll [14, 15], however, describes the process in two stages as an evaporation front moves into the coating leaving behind a "dry" layer, containing no continuous water, with ahead of it a transition layer, loosing water to the dry layer above and supplied with water from the wet latex below. The rapid rate is maintained as long as wet latex remains at the substrate, then the rate progressively declines.

The MFFT tends to be close to the glass transition temperature, (T_g), of the polymer and is reported for various polymers to be somewhat above or below the T_g [16, 17]. It is affected by the same molecular features, eg inclusion of a softer polymer will lower both the T_g and the MFFT. Ellgood [18] showed for a series of vinylidene chloride (VDC)/ ethyl acrylate (EA) copolymers that both T_g and MFFT peaked with an increasing VDC content, but not at the same composition. Below 55% VDC the T_g was greater than the MFFT, but above 55% the T_g was less than the MFFT with differences at the extremes of up to 15° C. Jensen [17] reports a lowering of the MFFT with decreasing particle size.

It is generally assumed that the latex particles have sufficient colloid stability to form a close packed array of particles prior to coalescence, but, for a soap-free latex, Okubo [19] reported a porous flocculated layer at the film interfaces. For a marginally stable latex, it was porous at the film-air interface and when deliberately destabilized further with added electrolyte then the substrate-film interface was also found to be porous. When stability was increased with added surfactant then a close packed, non-porous, structure resulted in agreement with Isaacs' findings [20].

Numerous theories have been suggested over the years concerning the fusion of latex spheres into polymer films, including:

i.   dry sintering [21, 22] driven by the polymer-air surface tension;

ii.   wet sintering [23, 24] driven by the polymer-water interfacial tension;

iii.   capillary pressure [25, 26] resulting from the negative curvature in the interstitial capillary system which develops between packed arrays of particles;

iv.   piston-like compression [13] which arises from a preferentially dried surface layer building in thickness from the top down;

v.   and interparticle cohesion promoted by surface forces [27].

There is still an active literature on the magnitude and relative importance of these forces which operate concurrently, as water evaporates, and tend to promote coalescence resisted by the viscoelasticity of the spheres. Greater complexity results when allowance is made for the changing area of contact upon drying and the effect of variable rates of drying in terms of the dependent changes of shear modulus and creep compliance [16, 24, 26, 28, 29]. According to the above theories, a film formed from a latex would comprise deformed polymer particles held together by physical forces. An important final stage of formation is when the film eventually develops its full strength by further gradual coalescence arising from autohesion [30, 31, 32, 33, 34], ie. the interdiffusion of chain ends between neighbouring particles.

Voyutskii [31] recognised that the dispersant moves away from the particle surfaces, dissolves in the polymer, (if compatible), or forms aggregates between particles, before polymer surfaces can contact and coalesce. Incompatible material may exude towards either film surface, as demonstrated by modern surface analytical techniques such as XPS, SIMS and FTIR [35, 36, 37].

That films become completely homogeneous after the coalescence process has been questioned by Kanig *et al* [38, 39, 40]. Previously, clear films tended to become turbid when swollen with water. Also, if monodisperse particles were involved, then Bragg scattering iridescence could be observed in the swollen film and if the film was stained and examined by transmission electron microscopy (TEM), then a honeycomb structure of interparticle boundaries could be demonstrated. Recent experimental advances [41, 42, 43] have enabled the study of the interdiffusion of polymer chains across interparticle boundaries. Progress in the understanding of interface healing [44] has been made which is also relevant to crack healing and polymer welding in terms of the de Gennes

[45, 46] reptation diffusion model or the biased reptation model involving charged end groups [45, 46]. SANS studies recently reviewed by Sperling [47] have shown that the full tensile strength is achieved in films when the penetration depth is equivalent to the radius of gyration for moderate molecular weights, when chain ends are located primarily on the particle surface. The theoretically predicted dependence of chain interdiffusion by reptation on time $t^{1/4}$ is obeyed following an induction period. Winnik [41] used fluorescence quenching techniques, with fluorescent dye in some particles, and acceptor in others, to study interdiffusion and found that for poly(butyl methacrylate), (PBMA), the further gradual coalescence followed the Williams-Landel-Ferry equation [48] with a temperature dependent activation energy. The diffusion coefficient, D, dropped dramatically with extent of penetration and to lower extents of penetration at lower temperatures. The apparent activation energy for diffusion was 160 kJ mol^{-1} at 51-124° C with initial diffusion coefficients ranging between 10^{-18} and 10^{-14} cm^{2}s^{-1}

Film permeation.

Permeation through non-porous polymers is explained [49, 50] in terms of solution and diffusion. The Flux, J, through unit area is given by:

$$J = \frac{D(C_1 - C_2)}{l} \qquad (2)$$

Where, D = Diffusion Coefficient;

l = thickness;

$C_1 - C_2$ is the concentration gradient within the film.

Also,

$$J = \frac{P(C_1' - C_2')}{l} \qquad (3)$$

Where, P = Permeability Coefficient;

$C_1' - C_2'$ = The concentration gradient in the external phase.

If Henry's Law is obeyed, then:

$$C = SC' \qquad (4)$$

where S = Solubility Coefficient.

Then:

$$P = DS \qquad (5)$$

Both D and S and hence also P can be concentration dependent. The activation energy for diffusion, (E_D), depends upon permeant size and level of plasticisation. For large molecules, E_D can be up to 160 kJ mol^{-1}, ie. of the same order as that for viscous flow in polymers [51]. The mechanism of diffusion [52] is the same both above and below the T_g, differing only in the frequency of motion of polymer segments and is satisfactorily described by Fujita's [50] "free volume theory", ie. molecules can only diffuse when the local free volume exceeds a critical value.

Partition of permeant into the polymer depends on solubility controlled by permeant-polymer interactions or more generally sorption which can include trapping in microvoids and cluster formation when permeant-permeant interactions are stronger than permeant-polymer interactions. Dual mode sorption considers Henry's Law sorption between polymer chains and Langmuir isotherm sorption in microvoids or on fillers [51].

The higher the density of the polymer, the lower is D, and increased rigidity of the polymer backbone leads to lower free volume whilst crystalline regions act as impermeable barriers.

A study of transport through polymer films can be directly relevant, (eg. where the films are to be used as barrier coatings or where they are used as controlled release coatings), or more indirectly relevant giving information on film morphology and changes in morphology.

Film preparation.

Yaseen [53] has reviewed a full range of film preparation techniques, and found pros and cons for each depending upon intended application. With surfactant-free latices there are problems in wetting the substrate and of film detachment compared with the surfactant-present case. Chainey *et al* [54] evaluated a number of methods such as coating on photographic paper and removal by soaking in warm water to dissolve the gelatin, or casting on mercury, silanised glass or PTFE, but rejected them on the grounds that either the substrate concerned contaminated the film or that the minimum thickness of film that it was possible to cast was at least an order of magnitude greater than required. Chainey [55] chose a "flash-casting" method with latex sprayed at a heated, PTFE coated, copper block at temperatures exceeding 120° C. Multi-passes, (20-50), of the spray gun were employed and the film was then allowed to return to room temperature or cooled to closer to the T_g of the polymer, using an appropriate carbon dioxide slush bath, before removal. Roulstone [56] cast poly(n-butyl methacrylate), (PBMA), films on Pyrex glass plates from which they could be removed by soaking in warm water. Roulstone also prepared *in situ* films on drug cores containing Ibuprofen, (2mm diameter beads with cellulose binder), using top spraying in a laboratory scale fluidised bed coater.

The aim of the following section is to review results obtained on the morphology and permeability of films prepared by the above three methods from "model colloid" latices. ie. made by surfactant-free emulsion polymerisation [57] monodisperse and typically ca 400 nm diameter, and cleaned by microfiltration [58]. The results are presented according to the methods of film preparation used.

Flash-cast free films.

Figure 1 shows a freshly cast PBMA latex film viewed by SEM. The surface of the film which was rough after casting, (figure 1a), smoothed out following one month storage at room temperature, (figure 1b), suggesting that a type of dry sintering mechanism was operating at the surface. In contrast freeze-fracture cross sections of the film interior showed no signs of structure or interparticle boundaries remaining, before or after aging even at the highest magnification employed [54].

Gas permeability coefficients were measured [59] by a dynamic technique in which helium in nitrogen was made to

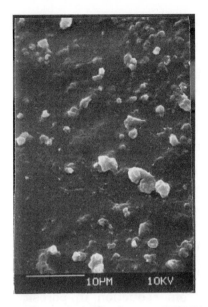

Figure 1 Flash cast PBMA latex film.
Figure 1a, (left), immediately after casting.
Figure 1b, (right), one month after casting.

flow over one side of the film and the helium which permeated
was swept away from the other side by a stream of pure
nitrogen which then passed to thermal conductivity detector.
A characteristic feature was that P was not constant but
decreased with time, (Figure 2), when monitored over some
800 hrs, falling to a near constant value still well above that,
$(1.16*10^{-16}$ s m^3 kg^{-1}), obtained for a solvent cast film of the
same polymer. For films cast at 438 K and stored at elevated
temperatures of 323 K & 353 K, curves of a similar shape

were obtained but the initial slopes were greater with an activation energy of 25 kJ mol^{-1}. A linear plot of P versus time, t, raised to ther power -0.25, (figure 3), compares favourably with the time dependence found for reptation model dynamics. Thus, although no evidence of remaining interparticle boundaries was detected by electron microscopy further densification of the polymer is implied by the permeability data.

Balik [60] has suggested that the large changes in permeability found by Chainey arise from a decreasing solubility of helium resulting from reduction in porosity, (microvoids, frozen in free volume, etc.), rather than solely changes in diffusivity of the gas. For a commercial terpolymer of vinyl chloride/ butyl acrylate/ vinyl acetate with a T_g of 1° C, Balik found a 5-fold change in CO_2 solubility at 30° C between fresh and stored latex films which first had displayed "dual mode" sorption and then the single mode Henry's Law solubility upon aging. The diffusivity was only 60% higher in the fresh film, but, with porosity well below 1% he pointed out that transport would still need to be via an interconnected pathway again suggestive of the importance of the interparticle boundary network.

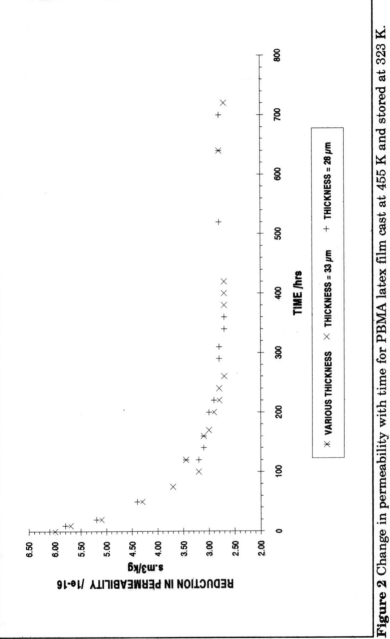

Figure 2 Change in permeability with time for PBMA latex film cast at 455 K and stored at 323 K.

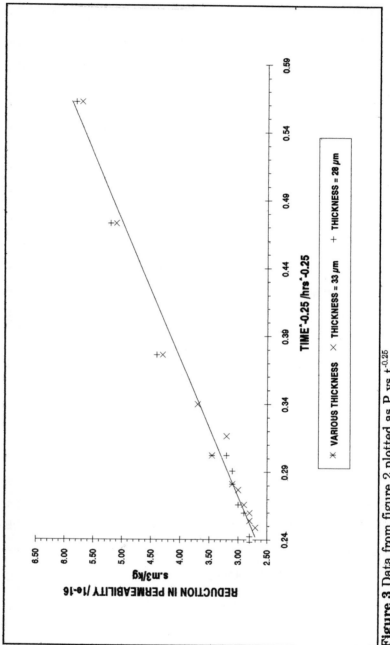

Figure 3 Data from figure 2 plotted as P vs t^{-0.25}

A reduction in the free volume, previously utilised by molecular motions, "frozen" in to the polymer films when cooled below the T_g has been discussed by List and others [61, 62, 63] in terms of "enthalpy relaxation" as the polymer chains in the metastable glassy state relax towards equilibrium.

Chainey [59] found that the permeabilities of solvent cast films prepared by the high temperature flash casting technique remained constant, however, over a period of several months. All the evidence points to the fact that the relaxations and densification in the latex films are associated with the interparticle boundary regions, even when they are sufficiently well coalesced not to be identified by SEM.

Films cast on Pyrex glass plates.

Here a lower casting temperature could be employed and, also in contrast to the spraying process described above, was not operator dependent involving only pouring and spreading. Surfactant-free latices do not wet low energy substrates, from which the film could be more easily removed when dry, sufficiently to produce thin films. On glass, problems arise from adhesion and films need to be soaked in water for detachment. Great care is needed to avoid damage by cracking or stretching of the film during removal.

PBMA films were cast [64] on level Pyrex glass plates constrained by Pyrex rings of 9.5 cm diameter in a laboratory oven, (no fan), containing a large vat of silica gel. Films cast at temperatures between 55° C and 95° C gave helium permeabilities greater than could be measured by the apparatus described previously or by the British Standard [65], (Daventest), method. Only when the films were further conditioned at 120° C, when considerable softening or partial melting occurred, did helium fluxes become measurable. The films were examined by a freeze-fracture TEM replication technique in which the sample was cooled by liquid nitrogen refrigerant during both the fracture and replication processes. For PBMA latex cast at room temperature, (figure 4), the structure is basically highly ordered over large areas. Hexagonal shapes, (A), can be seen corresponding to the cross-section of the original 400 nm latex particles fractured across their centres. The smooth diamond shapes, (B), correspond to the surfaces of particles in the layer below. Close packed spheres with a coordination number of 12, uniformly compressed to fill all space, became dodecahedral and the evidence here suggests that they are rhombic dodecahedra. These have two types of vertex, four of the trigonal based pyramid type and three of the square based pyramid type and evidence of these can be seen in figure 4. Joanicot [29] has recently demonstrated by SANS that rhombic dodecahedra

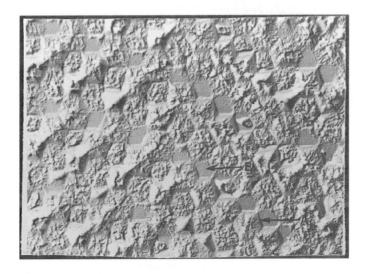

Figure 4 TEM replica of freeze-fracture cross section of PBMA film dish cast at room temperature.

result from face centred cubic close packed uniform spheres upon film formation.

For films prepared at 95° C interparticle boundaries are more coalesced, (figure 5), and this correlates with changes in permeability discussed below. Following aging at 120° C there is no longer any structural evidence of interparticle boundaries, (figure 6), and the fracture cross-section is very similar to that for a solvent cast film of the same polymer.

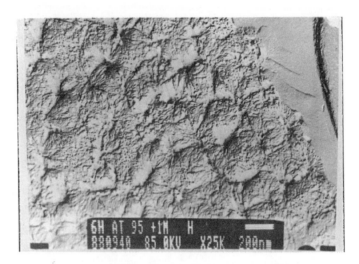

Figure 5 TEM replica of freeze-fracture cross section of PBMA latex film dried at 95° C for 6 hrs and stored 1 month at room temperature.

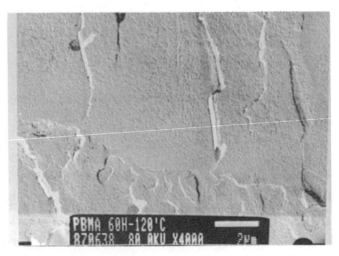

Figure 6 TEM replica of freeze-fracture cross section of PBMA latex film dried at 80° C for 3 hrs and aged at 120° C for 60 hrs.

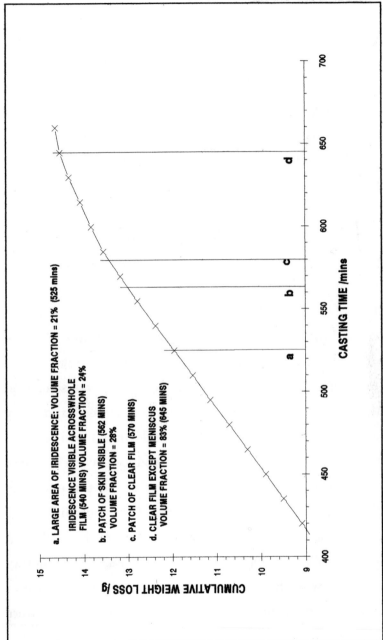

Figure 10 Viewed observations on PBMA latex film formation, (details in figure 9).

fraction, Φ, (as calculated from cumulative weight loss, as shown in figure 9), overall was 21%, then spread across the whole film surface. Then a skin became apparent first, at the site of the original iridescence, ($\Phi = 28\%$), and this became a patch of clear film, ($\Phi = 33\%$), with further skin around its edge which then spread to the whole film surface, ($\Phi = 83\%$), except for a wet meniscus, at the edge of the glass retaining ring.

As a consequence of the skin formation the upper surface of the final transparent film appears textured, (figure 11), with a more matt finish than the lower, substrate side, glossy surface. Okubo [19] suggested that skin formation was associated with instability in surfactant-free latices but here the PBMA latex had a critical coagulation concentration of 20 mM barium chloride. When the dried film was soaked in water, Bragg diffraction iridescence was observed but quickly disappeared, (a few seconds), upon air drying. It was restored on re-wetting. This effect could be reproduced over a period of two weeks but then ceased, again reflecting the effects of further gradual coalescence. Because of the side difference, due note was taken of the orientation of all of these films in mounting for permeability experiments. Permeability asymmetry was observed both for 4-nitrophenol and water vapour permeabilities. Water vapour permeabilities were

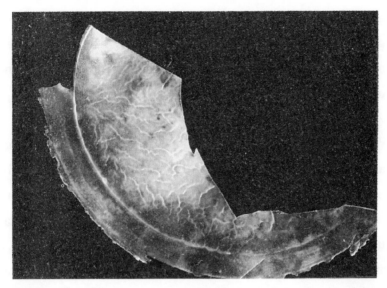

Figure 11 Transparent PBMA latex film cast at 40° C and side lit to show surface texture.

determined [68] for PBMA latex cast and solvent cast films, (freeze dried and dissolved in 2-butanone), of the same polymer, by a simple gravimetric method similar to that used by Banker [69]. 10 cm^3 sample bottles containing 5 cm^3 of saturated ammonium sulphate had films glued to their tops and were weighed as a function of time upon storage in a desiccator. Values, (substrate side towards the water vapour), for the 24 hrs old latex film and 1 month old film were 2.1($\pm$0.3) and 1.4($\pm$0.1) *10^{-17} s.m^{3}kg^{-1} respectively and for the solvent cast film 1.7($\pm$0.6) and 1.7($\pm$0.1) *10^{-17} s.m^{3}kg^{-1} respectively, for fresh and aged films. Thus, after aging the latex film actually has a lower permeability than the solvent cast film which showed no reduction in permeability upon aging. Latex films having lower permeabilities than their solvent cast equivalents has also been reported by List and Kassis [61].

When PBMA latices containing post added surfactant were cast on nylon plates, the films [68] could be removed from the substrate without soaking in water. Water vapour permeabilities were determined, and for sodium dodecyl sulphate, (SDS), addition it was found that there was a minimum in permeability at monolayer coverage which presumably can be attributed to better particle packing. Below monolayer coverage of SDS the films aged to lower values but above it did not age. Excess surfactant above monolayer coverage seems only to make the film more hydrophilic.With the non-ionic surfactant dodecyl tetraoxyethylene glycol monoether ($C_{12}E_4$), a minimum of permeability at monolayer coverage was again observed. Coalescence was increased but beyond this level of addition plasticisation increased permeability. In the presence of levels of the surfactant greater than monolayer coverage further gradual coalescence, to lower permeability, did not occur.

Bead coating in a fluidised bed.[1]

Spherical beads, (ca. 2 mm diameter), comprising Ibuprofen in a cellulose matrix were spray coated from above in a laboratory scale, (400 g batch), fluidised bead coater to give 10% by weight addition of PBMA latex. Spraying at temperatures of 38, 42, 50 and 60° C for 20 minutes was followed by drying in air at 45° C for 5 hrs.

Release profiles were obtained for 0.5 g samples of coated beads contained in gelatin capsules placed into wire baskets in a dissolution medium of phosphate buffer at 37° C and rotated at 100 rpm. The U.V. absorbance of this medium was detected in a spectrophotometer at 220 nm and six such samples were monitored simultaneously. The effects of aging, both at room temperature for 14 days and at 60° C for a further 2 days were determined.

It is apparent, (table 2), that the initial coating temperature could be used to control the initial release rate but that storage at room temperature reduced the release rate to a nearly constant value. Heat treatment at 60° C reduced the release rate further and makes the first order profile nearly

---

[1]The Authors are indebted to Dr. P. York, Postgraduate School of Pharmacy, University of Bradford, for provision of laboratory facilities and advice in carrying out the work.

Coat-ing temp. (°C)	First order releae rate constant /1*10⁻⁴ min⁻¹				
	5 hrs old	14 d @ room temp.	Extra 2 d @ 60° C	2 m'ths @ room temp.	extra 2 d @ 60° C
38	N/A	10.78	3.91	6.46	2.18
42	19.13	11.33	5.03	6.98	2.5
50	15.41	12.77	4.62	6.7	1.24
60	7.77	7.73	4.97	3.91	1.05

Table 2 The effect of temperature and storage on the release rate from beads coated at various temperatures.

linear, (figure 12). Although a linear release is desirable pharmaceutically the actual rate was too slow in this case with only 50% of the drug released in 24 hrs compared with a target value of nearer 100% in 12 hrs. Additional aging for 2 months at room temperature and heating to 60° C for 2 days demonstrated that further gradual coalescence was still possible with a further decline in release rate. As can be seen from the micrographs, the bead coats are cracked and porous initially, (figure 13). For a porous substrate the rate of water loss on film formation is affected by absorption of water into the substrate and this, together with surface irregularities, can lead to reduced film quality and crack formation [12]. It

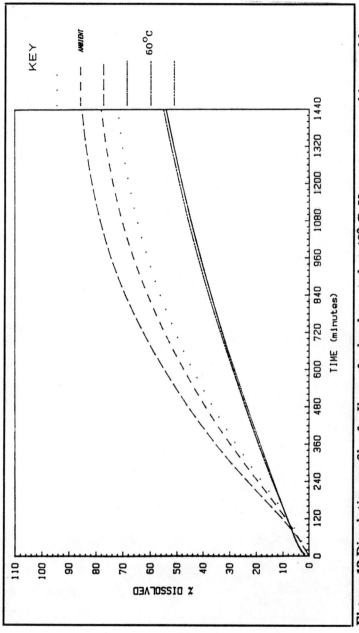

Figure 12 Dissolution profiles for Ibuprofen beads coated at 42° C. Upper curves ambient and lower aged at 60° C.

has been suggested that controlled drug release through a cracked porous coating is a viable drug release process [70]. Upon aging and heat treatment the beads, however, heal and even sintering between adjacent beads is observed, (figure 14). It is also apparent that some crystalline material has exuded to the surface of the bead coating, ie. the drug has been dissolved in water during spraying, been incorporated in the film coating, and has eventually exuded to the surface.

A combination of the porous and cracked nature of the coat initially, and the surface exudation of drug later, presumably explains why no time lag is observed prior to drug release. It is apparent here again, that for these additive-free latex

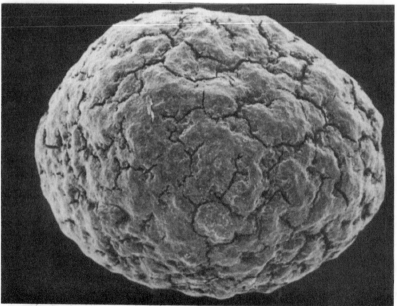

Figure 13 Beads coated with PBMA at 42° C.

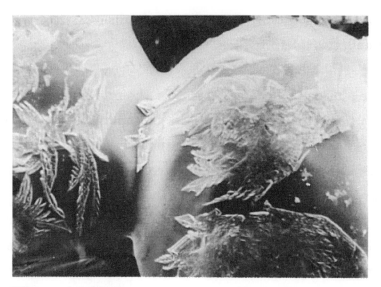

Figure 14 Beads coated with PBMA at 50° C, aged 2 months at ambient plus 2 days at 60° C.

coatings it is the further gradual coalescence process which hampers attempts at controlling drug release rate.

Conclusions.

Reported experimental verification of the relative importance of the various mechanisms described in the theories of latex film formation is very limited. Whilst different mechanisms operate to different extents under different preparation conditions, eg. whether the latex is sprayed or spread, and whether the substrate is impermeable or porous, it is the nonuniformity of drying, even in simple dish casting, which complicates interpretation.

Using "model colloid" surfactant-free homopolymer latices of high purity to prepare films has the advantage that their behaviour is not complicated by exudation of additives, or by their being trapped in the film matrix, or by their separation. Also, use of homopolymer latices avoids the problems of polymer compositional gradients or inhomogeneities in the latices or the films. This is, however, at the price of practical handling difficulties in terms of spreading and removal from a suitable substrate in order to produce a free film. Temperature and time are fundamental parameters for both the preparation of the films and their storage in relation to morphology and permeability characteristics. An understanding of these processes is important in attempting to control transport through polymer latex films, whether as a barrier coating or as a sustained release coating, where aqueous based formulations are to replace solvent based ones. The marked time and temperature densification of latex films can lead to the latter having lower permeability. Low levels of surfactant addition decreased latex film permeability even to water vapour and prevented aging for the PBMA latices employed in this study.

References.

1   Bindschaedler C., Gurny R., Doelker E., (1983). Labo-Pharma Probl. Tech., 31/331, 389-394.

2   Talen H.W. and Hover P.F., (1959). Deutsche Farben Z, 13, 50-55.

3   Napper D.H., (1969). J. Chem. Ed., 46, 305-307.

4   Funke W. and Zorll U., (1975). Defazet, 29, 146-153.

5   Nicholson J.W. and Wasson E.A., (1990). In Surface Coatings, Ch. 2, Vol. 3, 91-123.

6   Gould R.F., (1964). In Contact Angle, Wettabiliy and Adhesion, Advances in Chemistry, Series 43, ACS.

7   Kesting R.E., (1985). In Synthetic Polymeric Membranes:- A Structural Perspective, Ch. 4, 2nd Ed., 106-179.

8   Blandin H.P., David J.C., Vergnaud J.M., Illien J.P. and Malizewicz M., (1987). J. Coatings Tech., Vol. 59 (746), 27-32.

9   Vezin W.R., Florence A.T., (1981). European Polym. J., 17, 93-99.

10   Vanderhoff J.W., Bradford E.B. and Carrington W.K., (1973). J. Polym. Sci. Symp. 41, 155.

11   Pramojaney N., Poehlein G.W., and Vanderhoff J.W., (1980). Drying '80, 2, 93-100.

12   Bierwagen G.P., (1979). J. Coatings Tech., 51 (658), 117-126.

13   Sheetz D.P., (1965). J. Appl. Polym. Sci., 9, 3759-3773.

14   Croll S.G., (1986). J. Coatings Tech., 58 (734), 41-49.

15   Croll S.G., (1987). J. Coatings Tech., 59 (751), 81-92.

16   Eckersley S.T. and Rudin A., (1990). J. Coatings Tech., 62 (780), 89-100.

17   Jensen D.P. and Morgan L.W., (1991). J. Appl. Polym. Sci., 42, 2845-2849.

18   Ellgood B., (1985). JOCCA, 68, 164.

19   Okubo M., Takeya T., Tsutsumi Y., Kadooka T.,
      Matsumoto T., (1981). J. Polym. Sci. Polym. Chem. Ed.,
      19, 1-8.

20   Isaacs P.K., (1966). J. Macromol.Chem., 1 (1), 163-185.

21   Dillon R.E., Matheson L.A. and Bradford E.B., (1951).
      J. Colloid Sci., 6, 108.

22   Frenkel J., (1943). J. Phys., (USSR), 9, 385.

23   Vanderhoff J.W., Tarkowski H.L., Jenkins M.C. and
      Bradford E.B., (1966). J. Macromol.Chem., 1, 361.

24   Mason G., (1973). Brit Polym. J., 5, 101.

25   Brown G.L., (1956). J. Polymer Sci., 22, 423.

26   Lamprecht J., (1980). Colloid & Polym. Sci., 258 (8),
      960-967.

27   Kendall K. and Padget J.C., (1982). Int. J. Adhesion, 2,
      149.

28   Cansell F., Henry F. and Pichot C., (1990). J. Appl. Polym. Sci., 41, 547-563.

29   Joanicot M., Wong K., Maquet J., Chevalier Y., Pichot C., Graillet C., Lindner P., Rios L. and Cabane B., (1990). Prog. in Colloid & Polym. Sci., 81, 175-183.

30   Voyutskii S.S., (1958). J. Polymer Sci., 32, 528.

31   Voyutskii S.S., (1963). In <u>Autohesion and Adhesion of High Polymers</u>, Polymer Reviews, 4, Wiley International, New York.

32   Bradford E.B. and Vanderhoff J.W., (1966). J. Macromol. Chem., 1, 335.

33   Idem., (1972). J. Macromol. Sci. Phy., B6, 671.

34   Kast H., (1985). Macromol. Chem. Suppl. 10/11, 467.

35   Zhao C.L., Holl Y., Pith T. and Lambla M., (1987), Colloid and Polym. Sci., 265, 823-829.

36   Zhao C.L., Dobler F., Pith T., Holl Y., Lambla M., (1989). J. Colloid and Interface Sci., 128, 437.

37   Zhao C.L., Holl Y., Pith T. and Lambla M., (1989), Brit Polym. J., 21, 155.

38   Kanig G., Neff H., (1975). Colloid Polym. Sci., 253, 29.

39   Distler D., Kanig G., (1978). Colloid Polym. Sci., 256, 1052.

40   Distler D., Kanig G., (1980). Org. Coat. Plast. Chem., 43, 606.

41   Zhao C.L., Wang Y., Hruska Z. and Winnik M.A., (1990). Macromolecules, 23 (18), 4082-4087.

42   Hahn K., Ley G., Schuller H. and Oberthur R., (1986). Colloid Polym. Sci., 264, 1092.

43   Hahn K., Ley G., Oberthur R., (1988). Polymer Sci., 266, 631.

44   Kim Y.H. and Wool R.P., (1983). Macromolecules, 16, 1115-1120.

45   de Gennes P.G., (1971). J. Chem. Phys., 55, 572.

46   Slater G.W. and Noolandi J., (1986). Macromolecules, 19 (9), 2357.

47   Sperling L.H., Klein A., Yoo J.N., Kim K.O. and Mohammadi N., (1990). Polymers for Advnced Technologies, 1, 263-273.

48   Ferry J.D., (1980). In <u>Viscoelastic Properties of Polymers</u>, Wiley, New York.

49   Crank J., (1975). In <u>The Mathematics of Diffusion</u>, 2nd Ed., Oxford Univ. Press.

50   Crank J. and Park G.S., (Eds), (1968). In <u>Diffusion in Polymers</u>, Academic Press.

51   deV Naylor T., In <u>Comprehensive Polymer Science</u>, (Eds. Allen and Bebbington), Vol. 2, Polymer Properties, Ch. 20., Pergamon.

52   Rogers C.E., (1965). In <u>The Physics and Chemistry of the Organic Solid State</u>, (Eds. Fox D., Labes M.M. and Weissberger A.), Vol. 2, Cha. 6, 509, Wiley Interscience.

53   Yaseen M., Raju V.S.N., (1985). Prog. in Org. Coatings, 10, 125.

54   Chainey M., (1984). CNAA PhD Thesis, Trent Polytechnic.

55   Chainey M., Wilkinson M.C. and Hearn J., (1985). J. Appl. Polym. Sci., 30, 4273-4285.

56   Roulstone B.J., (1988). CNAA PhD Thesis, Trent Polytechnic.

57   Goodwin J.W., Hearn J., Ho C.C. and Ottewill R.H., (1974). Colloid and Polym. Sci., 252, 464-471.

58   Wilkinson M.C., Hearn J., Cope P. and Chainey M., (1981). Brit. Polym. J., 13, 82-89.

59   Chainey M., Wilkinson M.C. and Hearn J., (1985). J. Polym. Sci., Polym. Chem. Ed., 23, 2947-2972.

60   Balik C.M., Said M.A. and Hare T.M., (1989). J. Appl. Polym. Sci., 38, 557-569.

61   List P.H. and Kassis G., (1982). Acta Pharmaceutica Technologica, 28 (1).

62   Holsworth R.M., (1969). J. Paint Technol., 41, 167.

63   Illers K.H., (1969). Makromol. Chem., 127, 1.

64   Roulstone B.J, Wilkinson M.C., Hearn J and Wilson A.J., (1991). Polym. Int., 24, 87-94.

65   Daventest British Standard, 2782 part 8, method 821A.

66   Roulstone B.J., Hearn J., Wilkinson M.C., (1992). Polym. Int., 27 (1), 31-35.

67   Perera D.Y., (1985). JOCCA, 11, 275-281.

68   Roulstone B.J., Hearn J., Wilkinson M.C., (1992). Polym. Int. 27 (1), 43.

69   Banker G.S., Gore A.Y., Swarbrick J., (1966). J. Pharm. Pharmacol., 18, 457,

70   Ghebre-Sellassie I., Gordon R.H., Middleton D.L.,
      Nesbitt R.U. and Fawzi M.B., (1986). Int. J.
      Pharmaceutics, 31, 43-54.

MECHANICAL PROPERTIES OF COMPOSITE POLYMER LATEX FILMS

M. CHAINEY and P.A. REYNOLDS

Courtaulds Coatings, Stoneygate Lane, Felling,
Tyne and Wear, NE10 0JY

ABSTRACT

The film forming behaviour of polymer colloids depends, amongst other things, on the deformation of spherical particles under the influence of forces arising from the evaporation of the continuous aqueous phase. It is the purpose of this paper to discuss the film formation process in terms of the mechanical properties of the latex particles as inferred from measurements on latex films. A series of latices similar to that described by Lambla et al [*Makromol. Chem., Suppl.* 10/11, 436 (1985)] was prepared and characterised by dynamic mechanical analysis (DMA). Samples of films around 2 mm thick were formed by drying the latices in Petri dishes at $60^{\circ}C$. The films were compressed by a mass of up to 4 kg acting on a 15 mm diameter foot. The resulting vertical displacement was measured as a function of time. Subsequently, the recovery on removal of the weight was observed. Simple treatment available from squeeze-flow analysis allowed the viscosity of the thermoplastic polymer films to be calculated. The compression and recovery processes were treated as a creep compliance experiment. The results indicated that a film composed of core-shell latex particles had a higher viscosity than a gradient copolymer latex film of the same morphological sense. The recovery behaviour differentiated between the different types of morphology. The implications of these results on film formation from composite latex particles are discussed.

INTRODUCTION

The formation of films from polymer latices has been studied periodically since the early papers of Dillon et al (1) and Brown (2). The basic premise has been that, for film formation to occur, the forces exerted during the evaporation of water must exceed the resistance of the polymer to deformation. The nature of the forces promoting coalescence has been the topic of much debate, with one side favouring surface tension forces and the other capillary forces. A recent summary of the arguments has been presented by Eckersley and Rudin (3). Resistance to deformation is usually inferred from the modulus of the polymer as determined from dynamic mechanical measurements on bulk specimens. The minimum film formation temperature (MFT) of a latex is the temperature at which the modulus of the polymer is exceeded by the forces promoting coalescence. The MFT is often associated with the glass transition temperature of the polymer, and practically is observed within a few degrees of it (4). This situation is comprehensible with a dispersion of homogeneous particles, but it is less easy to visualise how a heterogeneous particle latex will behave on drying. Kast (5) has determined the modulus, as a function of temperature, and morphology of some composite latex films. The effect of surface hydrophilic material (e.g., copolymerised acrylic acid or poly(vinyl alcohol) stabiliser) on the final film properties was demonstrated. It was further shown that, in some cases, the morphology of a composite latex film could be modified by annealing, with a consequent change in modulus. Devon et al (6) have studied the film forming behaviour of a range of core-shell latex films, and concluded that where the shell thickness was sufficiently thin, the MFT was influenced by core as well as shell polymer properties.

The ultimate objective of the work reported in this paper is to understand the role of particle deformation in the film formation process, by inference from the cold flow of films. For the purposes of this preliminary study, an apparatus was required which was easy to construct, simple to operate, and which yielded data capable of straightforward interpretation. These requirements were satisfied by a dead weight tester (illustrated in Figure 1), in which films were compressed by a mass of up to 4 kg acting on a 15 mm diameter foot. The resulting vertical displacement was measured as a function of time. Subsequently, the recovery on removal of the weight could be observed. These conditions were expected to correspond to the region of the first Newtonian plateau. The experiment was treated as a creep compliance process. The expected result of such an experiment is illustrated in Figure 2 (7).

The squeeze flow experiment was carried out in three ways, as shown in Figure 3. In the first case, an expanse of film large by comparison with the dimensions of the foot was examined (the "infinite sea" experiment). In the second case, a disc of film smaller than the foot was examined, and in the third case, a similar disc with lubricated faces was investigated (8). It is inherently difficult to represent the exact viscometric flow for any of the cases, and therefore the analysis accepts a gross averaging of the flow processes.

In order that films of varying morphology could be examined, a series of latices similar to that described by Lambla et al (9) was prepared. The overall copolymer composition in each case was 55:45 methyl methacrylate (MMA)/butyl acrylate (BA) by weight. The uniform copolymer sample U1 was of this composition throughout. The core-shell samples C1 and C2 were respectively a 80:20 MMA/BA core/10:90 MMA/BA shell and vice versa, each having a core:shell ratio of 50:50. The gradient copolymer samples G1 and G2 had continuously varying copolymer compositions between the same extremes. Samples C1 and G1 had hard (high T_g) interiors and soft (low T_g) peripheries: samples C2 and G2 were of the opposite sense. Latex characteristics are listed in Table 1. Thin films cast from these latices were characterised by dynamic mechanical analysis (DMA), and thick films were used in the squeeze flow experiments.

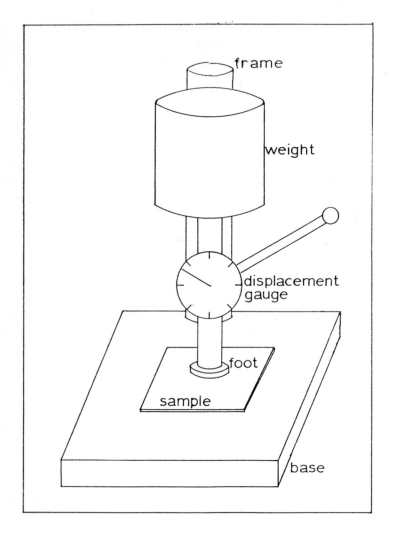

Figure 1. The dead weight tester used for the squeeze flow experiments.

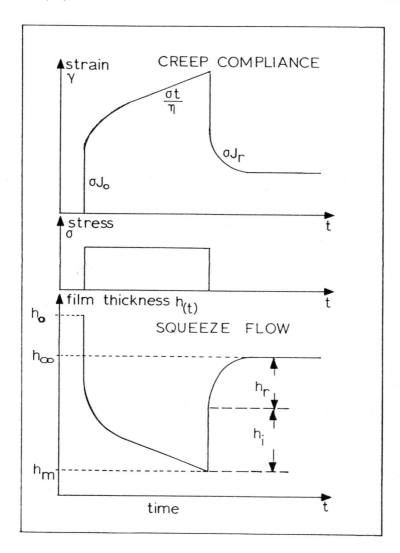

Figure 2. Comparison between the results from creep compliance and squeeze flow experiments.

TABLE 1

Details of Composite Latices

Sample	Morphology	Sense	$T_g/^oC$
U1	Uniform Copolymer		-3
C1	Core-shell	Hard/soft	57, -44
C2	Core-shell	Soft/hard	-44, 57
G1	Gradient copolymer	Hard/soft	Broad
G2	Gradient copolymer	Soft/hard	Broad

EXPERIMENTAL

Latex preparation

Uniform copolymer latex

This latex was prepared by semi-continuous emulsion polymerisation at 80°C, normally on a 2.5 kg scale at 45% NVC. A 3 dm^{-3} flanged reaction vessel, equipped with a condenser, nitrogen inlet, thermocouple and anchor stirrer was employed. The anionic surfactant (Gafac PE510) and most of the water were charged to the reactor, which was then placed in the water bath and allowed to attain the reaction temperature. The minimum quantity of concentrated ammonia solution required to dissolve the surfactant (2-3 cm^{-3}) was added, followed by 2% w/w of the monomer mixture. A solution of ammonium persulphate was added, to initiate the polymerisation. The reaction was left for 5 min. (seed stage), after which the remaining monomer was added at the required feed rate via a peristaltic pump. During the reaction, the exotherm was monitored: this was usually 2°C above the bath temperature. On completion of the feed, the reaction

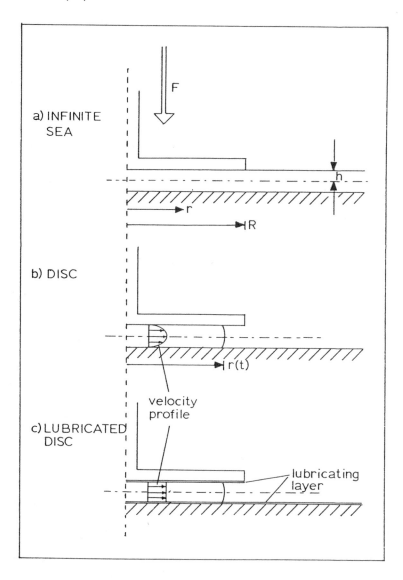

Figure 3. Comparison between the infinite sea, disc and lubricated disc experiments.

was left until the exotherm had decayed, after which the product was cooled and filtered. The particle size of the product was determined by photon correlation spectroscopy, using a Malvern model 4600 and making measurements at two angles. Non volatile content was determined gravimetrically.

Core-shell latex
Essentially, the same method was employed, except that two monomer mixtures of the appropriate compositions were added sequentially, with a 5 min. delay between the end of the first feed and the beginning of the second. At the end of the 5 min. delay, the exotherm was starting to decrease, indicating that the first monomer mixture had been almost completely consumed. The intention was to produce particles having a radial step change in composition.

Gradient copolymer latex
The power feed method of Bassett and Hoy (10) was employed. Again, two monomer mixtures were prepared, the first of which was contained in a stirred tank. The required amount of the first monomer mixture was added to the reactor for the seed stage. At the end of the seed stage, the contents of the stirred tank were fed to the reactor, and simultaneously, the second monomer mixture was fed into the first, at half the rate. Thus, a monomer mixture changing continuously from the first composition to the second was fed to the reactor, thereby producing particles having a radial gradient of composition.

Dynamic Mechanical Analysis

Measurements were made on the Rheometrics Solids Analyser (RSA II). Film samples were obtained by casting each latex on a glass panel with a 500 μm drawdown bar and allowing to dry overnight. Several specimens of dimensions 35 x 6 mm were cut from each plate. Three or four strips were mounted together in the instrument, to give sample dimensions of 24 x 6 x 1 mm. A standard 50 g pre-tension was applied to prevent the sample buckling during the experiment. Very soft films had to be cooled before applying the pre-tension. The instrument frequency was 1 Hz, and the temperature was increased from -50 to 150°C in steps of 3°C, allowing 25 s to attain thermal equilibrium before making a measurement. At 1 Hz, each measurement takes approximately 4 s. The results are presented as plots of tan δ against temperature.

Squeeze flow measurements

Squeeze flow measurements were made using a dead weight tester (Baty International Ltd., Burgess Hill, West Sussex, England. RH15 9LB). The apparatus comprised a 15 mm diameter foot on the bottom end of a vertical shaft. Weights of varying mass, to a maximum of 4 kg, could be mounted on the top of the shaft. The vertical displacement of the shaft was monitored by a dial gauge capable of measuring down to 1 μm. This assembly was mounted on a stand having a massive horizontal base. The foot was lapped so as to be parallel with the base. The dead weight tester was housed in a Perspex cabinet, within which the temperature was thermostatically controlled at 23 ± 1°C.

Films of around 2 mm thickness were prepared by drying latex at 60°C, i.e., above the T_g of any of the copolymers, in plastic Petri dishes or 50 mm diameter aluminium trays. These films were used directly in the infinite sea experiments. A few experiments were conducted on 10 mm discs cut from the cast films. The thicknesses of the films in the region where the foot was to be applied were accurately measured with a digital micrometer. In many cases, the dry films had many small bubbles of air trapped within them. As far as possible, areas of film free from bubbles were tested. These bubbles have been shown previously not to have a large effect on squeeze flow measurements (11). Films with grosser defects, such as ripples, fissures or substantially non-parallel faces, were rejected.

The films were placed under the foot of the dead weight tester. The shaft alone had a mass of 80 g, but this did not usually compress the film by a measurable amount. This mass is included in the stated masses with which the compressions were conducted. The weight was placed on the top of the shaft. This procedure took several seconds, a duration comparable with the timescale of the initial rapid compression. The whole procedure was recorded using a video camera and video cassette recorder, which superimposed the date and time on the image. After the compression experiment had run its course, the weight was removed, and the recovery recorded. This was necessarily conducted under the mass of the shaft, although this was small by comparison with the masses used in the compression experiments. The lubricated film experiment was conducted by placing a thin film of light silicone oil on the top and bottom faces of the latex film before applying the foot and the weight. With infinite sea films, the compression and recovery phases of the experiments were each run for 24 hours. With discs, the compression could only be continued while the edge of the disc remained inside the circumference of the foot, and thus the duration, and hence extent, of compression was very much less.

RESULTS

Latex characteristics

The latices were all reasonably monodisperse and had particle sizes around 100 nm. The formulation employed resulted in latices having low surface coverages of surfactant, and hence very little surfactant dissolved in the aqueous phase. As a result, migration of surfactant to the interfaces during film formation, with consequent lubricating effect in the squeeze flow experiment, was not expected to occur. DMA results, presented as plots of tan δ against temperature, are shown in Figure 4. These results are consistent with those reported by Lambla et al (9).

Effect of compression mass

Infinite sea film samples cast from latex U1 were compressed under three different masses, viz., 1.37, 2.47 and 4.00 kg (total mass, including that of the shaft), and then allowed to recover. The results, plotted as displacement, $h_{(t)}$, against time, are shown in Figure 5. The result for the 2.47 kg mass is apparently out of order because the film was significantly thinner than those used in the other two experiments. The data were treated using Stefan's equation (12,13). This relates the force, F, and the rate of displacement (dh/dt) for a non-lubricated film under constant area of compression at constant displacement rate, i.e.,

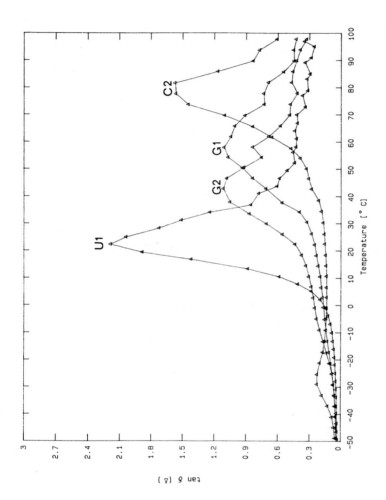

Figure 4. tan δ curves from dynamic mechanical analysis of the composite latex films.

$$F = \frac{3 \pi R^4 \eta}{8 h_{(t)}^3}\left(\frac{-dh}{dt}\right) \qquad (1)$$

where R is the radius of the foot and η is the viscosity of the film. For the experiments reported here, the form for a constant force and variable rate of displacement is appropriate (14), i.e.,

$$\left[\frac{1}{h_{(t)}^2} - \frac{1}{h_0^2}\right] = \frac{16Ft}{3\pi R^4 \eta} \qquad (2)$$

where h_0 is the half thickness at the start of the compression, and $h_{(t)}$ is the half thickness at time t. This treatment ignores inertial effects, normal force corrections and edge effects. For a Newtonian fluid, a plot of $[1/h_{(t)}^2 - 1/h_0^2]$ against time is expected to give a straight line of slope of $(16F/3\pi R^4)(1/\eta)$. Such a plot is shown in Figure 6. Values of viscosity calculated by this means are listed in Table 2. Agreement between the viscosities can be considered satisfactory, and suggests that zero shear conditions were close to having been achieved.

Effect of sample geometry

Samples of U1 latex film in the form of 10 mm diameter discs were compressed under the 4 kg mass and allowed to recover. In one case the faces of the film were lubricated with a light silicone oil. The results, together with that for an infinite sea experiment carried out using the same mass, are shown graphically in Figure 7.

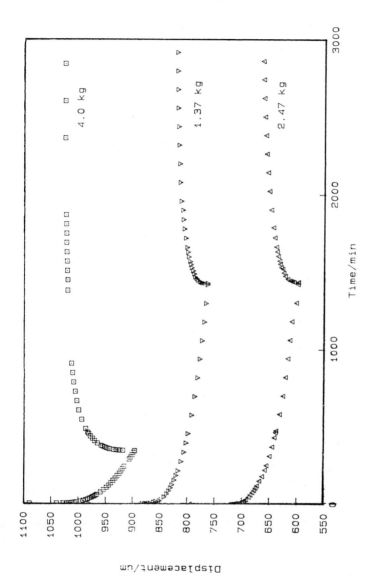

Figure 5. Displacement vs. time curves for squeeze flow experiments carried out with different masses.

TABLE 2

Effect of Compression Mass on Latex U1 Film Behaviour

Mass/kg	Film thickness /mm	10^{-9} Viscosity /Pa s	10^6 Shear Rate /s^{-1}
1.37	2.178	2.33	3.3
2.47	1.590	3.15	4.4
4.00	1.960	2.45	9.2

Effect of particle morphology

Infinite sea film samples cast from the gradient copolymer and core-shell latices were compressed under the 4 kg mass and allowed to recover. Viscosities calculated from equation (2) are listed in Table 3.

DISCUSSION

Squeeze flow as a creep compliance experiment

The results described above are very much those expected of a creep compliance experiment (7). The linear plots of $[1/h_{(t)}^2 - 1/h_0^2]$ against time give calculated values of viscosity of reasonable magnitude. Confidence in this method for determining viscosity is enhanced by the reproducibility of the results obtained with films of differing thickness cast from the same latex. Estimated errors of about $\pm$ 25% appear reasonable, given that no corrections have been calculated for inertia, normal force (15) and edge effects.

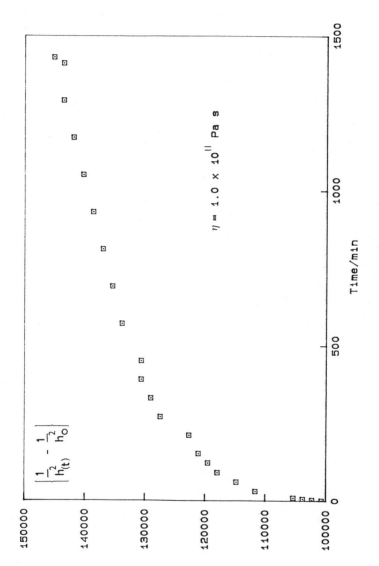

Figure 6. Plot showing calculation of viscosity from equation 2.

TABLE 3

Effect of Morphology

Sample	Morphology Sense	Viscosity /Pa s	Recovery h_i	τ
U1	Uniform	2.45×10^9	0.19	medium
C1	Core-shell Hard/soft	4.97×10^{11}	0.41	high
C2	Core-shell Soft/hard	1.00×10^{11}	0.52	high
G1	Gradient Hard/soft	1.69×10^{11}	0	low
G2	Gradient Soft/hard	6.67×10^9	0	low

Moreover, the data analysis does not account for the complex flow behaviour under the area of compression (16), but considers only a gross average of these flows. The non-lubricated squeezing flow has components of both shear and biaxial extension, and the biaxial extensional viscosity is expected to be six times that of the shear viscosity for a Newtonian fluid (8,17).

The plots of $h_{(t)}$ against t tend to show that the initial compliance represented by the large drop at the beginning of the compression experiment is not recovered when the mass is removed. In part, this may be due to the faces of the film sample not being perfectly parallel. As a result, the first part of the compression takes place under the influence of the weight acting on an increasing area of the foot, until the point where the whole area is in contact with the film. It is also probable that the absence of the recovered instantaneous compliance results from the long duration of the compression which takes the experimental conditions outside the linear viscoelastic region. This departure from the linear region cannot be very great, however, because a significant proportion of the delayed elastic response is observed reproducibly in most of the experiments. Furthermore, the values of viscosity calculated from compression experiments carried out with different masses are similar. The calculated viscosities may not be the limiting zero shear values, but they are not expected to differ significantly, given the very low shear rates typical of the squeeze flow method as used here. Previous work by McClelland and Finlayson (11) on the squeeze flow of highly viscous polymers demonstrated that at low squeezing rates, the polymers behaved as Newtonian fluids. However, the molecular weights were of the order of 10^3, as opposed to 10^5 - 10^6 which is typical of latex polymer.

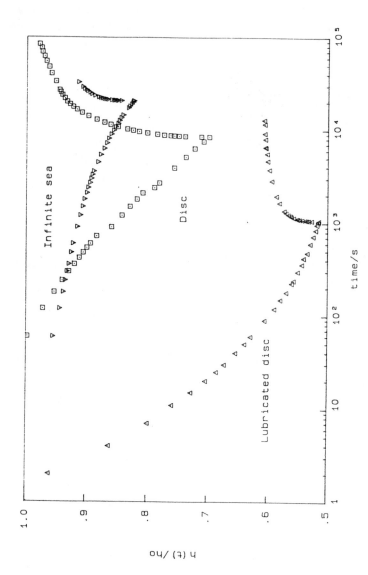

Figure 7. Comparison of squeeze flow data for infinite sea, disc and lubricated disc experiments on latex U1

Interpretation of data on composite latex films

The squeeze flow technique is clearly able to distinguish between the various types of morphology, although the reasons for the observed behaviour are not clear. On the assumption that it is the nature of the shell polymer which determines the flow behaviour, the viscosities would be expected to vary in the order

$$C2 > G2 > U1 > G1 > C1$$

since the viscosity ought to be reflected in the T_g relative to the temperature of the experiment. Instead, the observed order is

$$C1 > G1 > C2 \gg G2 > U1$$

This is obviously due to influence of the core polymer. The 50:50 core/shell volume ratio, and an ultimate particle size of around 100 nm, gives a calculated shell thickness of 10 nm, which is clearly too thin to screen the core (6). While there is no step change in composition in the gradient copolymer latex particles, there must be a radial discontinuity within the stressed particle where the local response changes from core to shell behaviour. Presumably the position of the radial discontinuity changes with temperature relative to the T_g range in the particle. That the uniform copolymer morphology exhibits the lowest viscosity must be due to the absence of polymer which is glassy at the temperature of the experiment. It therefore follows that in the composite particle films, the flow is considerably hindered by the presence of the glassy component.

Retardation Behaviour

In order to compare the retardation behaviour of the composite latex films used here, the data were plotted as $(h_{(t)} - h_m)/(h_0 - h_m)$ against t (Figure 8). This plot reveals a striking difference between the behaviour of the gradient copolymer and the core-shell latex films.

The data presented here show that the recovery which occurs on removal of the stress is an important facet of the mechanical behaviour of the film. The retardation effects occurring after the compression phase of a squeeze

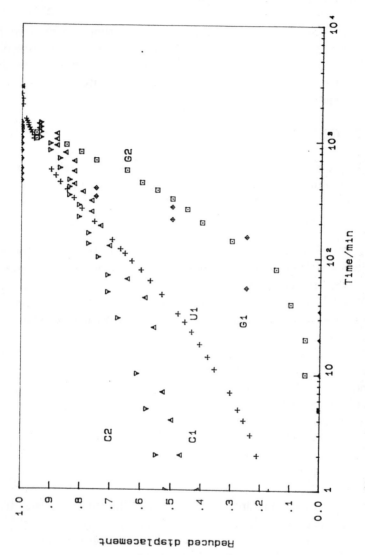

Figure 8. Strain recovery behaviour of composite latex films

flow experiment have seldom been studied. However, a recent paper (18) reports the study of the stress relaxation of a Bingham fluid after squeeze flow. The relaxation was fitted to exponential decays (stretched exponentials and the sum of two exponentials). The characteristic times (scaling of the exponents) were descriptive of the physical properties of the material, i.e., the concentration of bentonite in a clay based fluid.

The creep compliance of a Burgess body can be described by the expression (7)

$$J_{(t)} = J_0 + \frac{t}{\eta} + J_r\left(1 - e^{-t/\tau}\right) \tag{3}$$

where the compliance $J_{(t)} = (\gamma/\sigma)$ has an instantaneous component J_0, and a retarded component J_r which has a characteristic retardation time τ. This is shown schematically in Figure 2. If we consider the retardation spectra of the composite latex films (Figure 8), by relating the recovered height to the compliance we could by comparison assume a relation of the form

$$h_{(t)} = h_i + h_r\left(1 - e^{-t/\tau}\right) \tag{4}$$

where h_i is the instantaneous recovery, h_r is the retarded response, and τ is the characteristic time. Using the reduced form of the displacement (Figure 8), values of h_i have been assigned to the composite latex films (Table 3). The retardation behaviour does not fit well to a simple exponential form, as described by equation 4, as is evident from the non-linear nature of the plots in Figure 8. For this reason only a comparative description of the characteristic time is given. It is not surprising that the recovery behaviour is not characterised by a single retardation time since, with the exception of latex U1, the particles are expected to behave as though they are composed of two distinct phases, under the conditions of the experiment.

The (small) rapid response of the delayed recovery, and the absence of an instantaneous response, observed with the gradient copolymer latex films contrasts strikingly with the large instantaneous recovery and long delayed response of the core-shell latex films. Unlike the viscosity results, the retardation behaviour reveals similarities between films cast from latex of the same morphology, irrespective of the sense of the components.

Refinements to the Squeeze Flow Technique

The work described in this paper shows that valuable viscoelastic information is available from the squeeze flow technique. However, the foregoing discussion suggests modifications to both the experimental method and the analysis of the squeeze flow behaviour as a creep compliance. The most obvious improvement in experimental technique would be the production of film samples having parallel faces. The sample could be subjected to a small compression in order to produce the required parallel geometry. Removal of the compressing mass whilst keeping the foot in contact with the sample should retain the sample in a condition where a fully quantitative experiment can commence. The compression can then commence without removing the sample. Accurately linear viscoelastic behaviour in infinite sea experiments may be determined by curtailing the duration of the compression, thereby remaining within the limits of the linear viscoelastic region. This would reveal whether a full recovery of the instantaneous elastic response is obtained, and would also allow the zero shear viscosity to be calculated.

The infinite sea experiment is preferred on the grounds of simplicity. The use of a disc or lubricated disc really requires that $r_{(t)}$ be monitored, something which is not easily accomplished. It is possible to calculate $r_{(t)}$, assuming the sample density remains constant, from the equation (19)

$$r_{(t)} = r_0 \sqrt{h_0 / h_{(t)}} \tag{5}$$

A further disadvantage with the use of discs is the much greater effect of a slightly non-parallel geometry.

CONCLUSIONS

The squeeze flow experiment has given valuable viscoelastic information when the data are analysed as a creep compliance process. A complete interpretation of the results in terms of the latex particle morphology has not been possible, but some general tendencies can be seen.

From the compression phase of the experiment, the viscosity of films cast from core-shell latices is higher than gradient copolymer latices of the same morphological sense. For a particular morphology, films cast from latex particles having a soft periphery have a higher viscosity than films cast from latex particles having a hard periphery. The lowest viscosity is observed with films cast from the uniform copolymer latex.

On removal of the mass, the instantaneous recoverable strain from gradient copolymer latex films is zero, and the delayed time recoverable strain is small but relatively rapid. Conversely, the instantaneous recoverable strain from core-shell latex films is high, and the delayed time recoverable strain is relatively slow. For the uniform copolymer latex films, the instantaneous recoverable strain is high, and the delayed time recoverable strain is intermediate between the gradient copolymer and core-shell latex films.

REFERENCES

1. R.E. Dillon, L.A. Matheson and E.B. Bradford (1951), *J. Colloid Sci.*, **6**, 108

2. G.L. Brown (1956), *J. Polym. Sci.*, **22**, 423

3. S.T. Eckersley and A. Rudin (1990), *J. Coatings Tech.*, **62**, 89

4. J.G. Brodnyan and T. Konen (1964), *J. Appl. Polym. Sci.*, **8**, 687

5. H. Kast (1985), *Makromol. Chem., Suppl.* 10/11, 447

6. M.J. Devon, J.L. Gardon, G. Roberts and A. Rudin (1990), *J. Appl. Polym. Sci.*, **39**, 2119

7. G. Harrison (1976), *The Dynamic Properties of Supercooled Liquids*, Academic Press

8. S.H. Chatraei, C.W. Macosko and H.H. Winter (1981), *J. Rheology*, **25**, 433

9. M. Lambla, B. Schlund, E. Lazarus and T. Pith (1985), *Makromol. Chem., Suppl.* 10/11, 436

10. D.R. Bassett and K.L. Hoy (1981), in *Emulsion Polymers and Emulsion Polymerization*, D.R.Bassett and A.E. Hamielec (eds.), ACS Symposium Series No. 165, Chapter 23, p.371

11. M.A. McClelland and B.A. Finlayson (1988), *J. Rheology*, **32**, 101

12. J. Stefan (1874), *Akad. Wiss. Math. Natur., Wien*, **69**, Part 2, 713

13. O. Reynolds (1886), *Phil. Trans. Roy. Soc.*, **A177**, 157

14. R.B. Bird, R.C. Armstrong and O. Hassager (1987), *Dynamics of Polymeric Liquids*, Volume 1, Fluid Mechanics, 2nd ed., Wiley Interscience, New York.

15. G. Winther, K. Almdal and O. Kramer (1991), *J. Non-Newtonian Fluid Mech.*, **39**, 119

16. N. Phan-Thien, F. Sugeng and R.I. Tanner (1987), *J. Non-Newtonian Fluid Mech.*, **24**, 97

17. T.C. Hsu and I.R Harrison (1991), *Polym. Eng. Sci.*, **31**, 223

18. J.D. Sherwood, G.H. Meeten, C.A. Farrow and N.J. Alderman (1991), *J. Non-Newtonian Fluid Mech.*, **39**, 311

19. P.R. Soskey and H.H. Winter (1985), *J. Rheology*, **29**, 493

Acknowledgement

The authors are grateful to Dr. Mark Paterson and Martyn Care for the DMA measurements.

CONDUCTING POLYMER COLLOIDS - A REVIEW

S. P. Armes, School of Chemistry and Molecular Sciences,
University of Sussex, Falmer, Brighton, BN1 9QJ, U.K.

Abstract

The preparation and characterisation of sterically-stabilised colloidal dispersions of intrinsically conducting polymers is reviewed. We outline various dispersion polymerisation syntheses of polypyrrole and polyaniline in both aqueous and non-aqueous media. The application of a wide range of experimental techniques for the characterisation of such dispersions is discussed, including electron microscopy, CHN microanalyses, visible absorption, [1]H n.m.r. and vibrational spectroscopy, photon correlation spectroscopy, scanning-tunneling microscopy charge-velocity analysis, small-angle X-ray scattering, disc centrifuge photosedimentometry and surface-enhanced Raman spectroscopy. In particular, we consider the influence of the nature of the steric stabiliser on the physical and chemical properties of the conducting polymer dispersions (*e.g.* particle size and morphology, stabiliser/conducting polymer mass ratio, and electrical conductivity). Finally, various potential applications of these conducting polymer colloids are discussed.

Introduction

There has been enormous world-wide interest in organic conducting polymers since the discovery of highly conducting iodine-doped polyacetylene films by MacDiarmid's group in 1977 [1]. Unfortunately the highly conjugated molecular structure in these materials (a prerequisite for their metallic-like conductivity) results in the individual polymer chains being stiff and inflexible. As a consequence, conducting polymers are usually prepared as intractable films, gels or powders which are insoluble in common solvents and which decompose without melting when heated. In recent years considerable research activity has focused on developing novel *processable* forms of relatively air-stable conducting polymers such as polypyrrole and polyaniline. Many groups have synthesised soluble polypyrrole and polyaniline derivatives [2-6]. An alternative approach is the preparation of *colloidal dispersions* of such materials. There have been several reports in the literature concerning the preparation of microscopic composites of conducting polymers with preformed colloidal particles. For example, Yassar *et al.* have described the uniform coating of functionalised polystyrene latex particles with polypyrrole *via* an *in situ* deposition process [7a], whilst Partch *et al.* have utilised inorganic oxide particles which contain transition metal oxidants and so the pyrrole monomer is polymerised *exclusively* at the surface of the particles [7b]. However, most workers have focused on dispersion polymerisation techniques as the preferred method of synthesising conducting polymer colloids [8-35]. In this approach the conducting polymer is synthesised in the presence of a suitable polymeric surfactant or steric stabiliser. This stabiliser adsorbs onto the growing microscopic conducting polymer nuclei (*via* either physical adsorption in the case of polypyrrole or chemical grafting in the case of polyaniline) and prevents the usual macroscopic precipitation of the particles by a 'steric stabilisation' mechanism [36]. The result is a stable colloidal dispersion of conducting polymer particles (see Fig. 1). Clearly the stabiliser is the key component in such syntheses. We have generally found that the chemical structure of the stabiliser is the most important factor in determining the particle size of polypyrrole particles prepared in aqueous media [16]. Similarly, Vincent and Waterson have shown that the type of stabiliser can influence the particle morphology of polyaniline dispersions [29]. Other physical properties of these conducting polymer dispersions also depend on the type of stabiliser. For example, polypyrrole colloids stabilised with methyl(cellulose) or poly(vinyl methyl ether) flocculate reversibly at elevated temperatures due to the inverse temperature-solubility properties of the respective polymeric stabilisers [8,34]. In addition, Armes *et al.* have shown that poly(vinyl pyridine)-based stabilisers exhibit pH-dependent flocculation/stabilisation behaviour due to reversible protonation of the stabiliser's pyridine groups [11,12].

In the last five years polypyrrole and polyaniline colloids have been prepared and characterised in aqueous and non-aqueous media utilising a wide range of polymeric [8-35] and, more recently, particulate dispersants [37-39]. In the present work we review this conducting polymer colloid literature and examine the influence of the

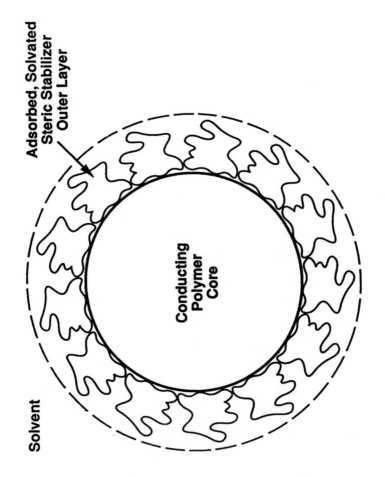

Fig. 1: Schematic representation of an isolated sterically–
stabilised conducting polymer particle.

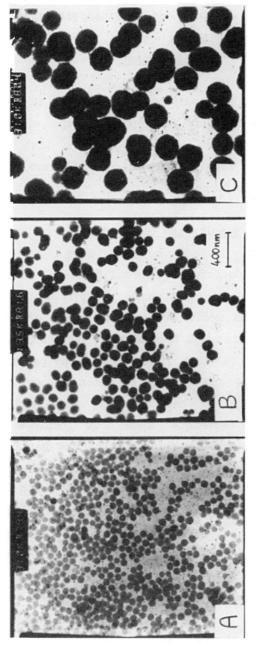

Fig. 2: Transmission electron micrographs of polypyrrole colloids prepared in aqueous media using different steric stabilisers: (a) PVA(125,000); (b) P2VP-BM and (c) PEO(700,00). Reproduced from Armes *et al., J. Colloid Interface Sci.,* 118, 410 (1987).

dispersant on the chemical and physical properties of the conducting polymer particles.

Experimental

All experimental procedures are described in detail by the authors in their original articles and will not be discussed further here. An explanatory list of the abbreviations used for the polymeric stabilisers in Tables 1-3 is given in the Appendix.

Results and Discussion

(1)   Polypyrrole Colloids in Aqueous Media:
Stable colloidal dispersions of polypyrrole are easily prepared in aqueous media with a wide range of polymeric stabilisers (see Table 1). Transmission electron microscopy studies have confirmed that in most cases the particles are generally spherical and relatively monodisperse ($\pm$ 15 % standard deviation). We have recently utilised a novel particle sizing technique known as 'charge-velocity analysis' (CVA) to accurately determine the particle size distributions of such dispersions [16]. The average particle diameter of the polypyrrole core can be varied over the range 66-300 nm simply by using an appropriate stabiliser (see Fig. 2). The particle size is also affected to a rather smaller extent by the initial stabiliser concentration and the stabiliser molecular weight. Usually statistical copolymers such as poly(vinyl alcohol-co-vinyl acetate) [9,10] or poly(4- or 2-vinyl pyridine-co-*n*-butyl methacrylate) [11,12] are used as steric stabilisers but

Table 1: Effect of steric stabiliser on particle size,
stabiliser/polypyrrole mass ratio and conductivity for
polypyrrole colloids prepared in aqueous media.

Stabiliser Type (MW)	Stabiliser/ polypyrrole mass ratio	Particle Size (nm)	Solid-state Conductivity (S cm^{-1})	Ref
MC(95,000)	1:9	100-200	0.2	7
PNVP(360,000)	?	70-150	0.14	7,9
PVA(25,000)	1:4	66+8	0.03	16
PVA(125,000)	1:10	89+15	0.5-1.7	10,16
P4VP(218,000)	?	150	?	11
P4VP-BM	1:4	91+17	0.4-2.0	11,16
P2VP-BM	1:5	130-200*	1.6-2.2	12
PEO(>10^5)	1:33	297+57	0.002	16,24

* depending on the initial stabiliser concentration.

several homopolymers such as poly(ethylene oxide) [16,24],
poly(N-vinyl pyrrolidone) [8,9] and poly(vinyl methyl
ether) have also been reported [34]. Very recently we have
shown that a tailor-made poly(N,N' dimethylaminoethyl
methacrylate-b-*n*-butyl methacrylate) block copolymer
stabiliser of narrow molecular weight distribution is also
an effective stabiliser for polypyrrole colloids [35].

We have utilised various analytical and/or spectroscopic
techniques in order to determine the stabiliser/conducting
polymer mass ratio in these sterically stabilised

polypyrrole colloids. For example, if the stabiliser has a zero nitrogen content (*e.g.* poly(vinyl methyl ether) [34]) then the stabiliser/conducting polymer mass ratio may be calculated *directly* from the reduced nitrogen content of the dried colloid relative to that of bulk polypyrrole powder (*ca.* 15.5%). Alternatively, for stabilisers such as poly(2- or 4-vinyl pyridine-co-*n*-butyl methacrylate) which have nitrogen contents similar to polypyrrole, the stabiliser/conducting polymer mass ratio may be estimated *indirectly* by analysing the post-reaction supernatant for non-adsorbed stabiliser by some convenient spectroscopic method such as Raman or F.T.I.R. spectroscopy [11,12]. Thus the adsorbed mass of stabiliser is easily calculated from the difference between the initial (known) stabiliser concentration and the final (measured) stabiliser concentration. Colorimetry techniques can also be utilised to assay post-reaction supernatant solutions [10]. In general we find that the stabiliser/conducting polymer mass ratio for polypyrrole colloids usually lies in the range 1:3 to 1:7 (*i.e.* the particles comprise 7-25 % stabiliser on a mass basis).

We have already seen that electron microscopy is a powerful technique for the elucidation of the particle morphology of diluted, dried-down polypyrrole colloids. Armes *et al.* have reported that scanning-tunneling microscopy can be utilised to examine the nanomorphology of individual conducting polymer particles [40]. Colloidal particles of both polypyrrole and polyaniline were observed to be made up of clusters of much smaller 'nanoparticulates'. Rawi *et al.* have examined the dimensions of the solvated, adsorbed stabiliser layer of poly(4-vinylpyridine-co-n-butylmethacrylate)-stabilised polypyrrole particles using a combination of photon correlation spectroscopy and transmission electron microscopy [22]. These workers estimated an upper limit stabiliser layer thickness of *ca.* 25 nm.

Epron *et al.* have measured the complex conductivity of methyl(cellulose)-stabilised polypyrrole colloids in their *dispersed, colloidal state* by microwave measurements [30]. These conductivities are approximately an order of magnitude lower than that of bulk polypyrrole powder (ca. 10 S cm^{-1}) and are in good agreement with the solid-state macroscopic conductivities observed for thin films or pellets fabricated from such colloidal dispersions. Thus these results suggest that the intrinsic conductivity of the colloidal polypyrrole particles may not actually be reduced by their electrically insulating stabiliser component. We note that Nimtz *et al.* have recently claimed that colloidal metallic particles exhibit appreciably lower conductivities than "macroscopic" metal samples simply by virtue of the sub-micron dimensions of the former system [41].

In collaboration with Dr. S. Y. Luk of Courtaulds Research we are currently using surface-enhanced Raman spectroscopy (SERS) in order to probe the structure of the adsorbed stabiliser layer of both polypyrrole and polyaniline particles in the solid state at the molecular level [42]. Our initial results (obtained on dispersions spin-coated onto gold substrates) show a SERS effect for some of the conducting polymer vibrational modes. This means that the conducting polymer component of the particles lies within ca. 10 Å of the gold surface and suggests that the outer layer of adsorbed stabiliser is relatively non-uniform and has 'bare' (or at least very thin) patches. Thus the conducting polymer particle cores are not effectively isolated in the solid state and probably make direct physical contact with each other. These results are fully consistent with the observed macroscopic conductivities of these materials.

(2)   Polypyrrole Colloids in Non-aqueous Media:

In 1986 Myers showed that pyrrole could be polymerised by $FeCl_3$ in several non-aqueous solvents such as diethyl ether or ethyl acetate to give highly conductive polypyrrole bulk powders [43]. Although it has been known for some time that polypyrrole colloids prepared in *aqueous* media could be redispersed in water-miscible solvents [9a,24], it is only relatively recently that Armes and Aldissi reported the synthesis of sterically-stabilised polypyrrole colloids directly in non-aqueous solvents [13]. Since this initial report there have been several publications describing the preparation of colloidal polypyrrole in various solvent systems (see Table 2).

In all cases the polypyrrole particles are spherical and rather more polydisperse than those prepared in aqueous media ($\pm$ 40 % *vs.* $\pm$ 15 % standard deviation as measured by charge-velocity analysis [13,16]). The reason for this increased polydispersity is not understood at present but this author believes that it is probably a result of the faster (and hence less controlled) pyrrole polymerisation

Table 2: Effect of steric stabiliser on particle size range and conductivity for polypyrrole colloids prepared in various non-aqueous solvents.

Stabiliser Type	Solvent	Particle Size Range (nm)	Solid-state Conductivity (S cm^{-1})	Ref
PVAc	Methyl Acetate	225$\pm$75	0.1	13
PVAc	Methyl Formate	276$\pm$90	0.06	13
PVAc	Ethyl Formate	?	?	13
PVAc	Propyl Formate	259$\pm$82	7.5 x 10^{-6}	13
PVAc	2-Methoxyethanol[a]	30-100	1.5 x 10^{-3}	21
PEO	Methanol	100-250	?	31
PVME	Ethanol/water[b]	50-150	1-2	34

a. Stabiliser/polypyrrole mass ratio = 1:7

b. Stabiliser/polypyrrole mass ratio = 1:6 to 1:16 depending on the initial stabiliser concentration)

kinetics observed for non-aqueous solvents. The solid-state conductivities of thin films or compressed pellets of non-aqueous polypyrrole dispersions are rather variable (10^{-5} to 0.2 S cm^{-1}). The reason(s) for the lower values are not known at present.

We have prepared polypyrrole colloids using poly(vinyl acetate) as a steric stabiliser in several alkyl formate solvents [13], methyl acetate [13], 2-methoxyethanol [21] and methanol [44]. In principle the stabiliser/polypyrrole mass ratio of such dispersions can be determined directly by nitrogen microanalysis. In practice, however, we have found this method to be rather unreliable due to the widely differing iron content (in the form of dopant anions and/or physically incorporated species) of such colloid samples relative to polypyrrole bulk powder 'control' samples. For polypyrrole dispersions prepared in 2-methoxyethanol solutions we have assayed the post-reaction supernatant solution for non-adsorbed poly(vinyl acetate) by F.T.I.R. spectroscopy. These results suggest a stabiliser/ polypyrrole mass ratio of approximately 1:7.

Very recently Odegard et al. have described the use of poly(ethylene oxide) as a stabiliser for polypyrrole prepared in methanol [31] and Digar et al. have reported that poly(vinyl methyl ether)-stabilised polypyrrole dispersions can be prepared in a 50/50 ethanol/water solvent mixture [34]. This latter system is particularly interesting because the authors have shown that such particles exhibit thermally-induced reversible flocculation behaviour and, in addition, can be easily dispersed in a polystyrene matrix. The latter observation is rationalised in terms of the well-known miscibility of poly(vinyl methyl ether) and polystyrene.

Finally we note in passing that Whang et al. have reported remarkably high conductivities for polypyrrole bulk powders prepared in 86/14 v/v acetonitrile/methanol solvent mixtures [45]. It would clearly be of interest to prepare sterically-stabilised polypyrrole colloids in this mixed solvent medium in order to try to improve the conductivity of such non-aqueous dispersions.

(3)   Polyaniline Colloids in Aqueous Media:

The Bristol group initially attempted to prepare sterically-stabilised polyaniline colloids using the same steric stabilisers as used for polypyrrole dispersions. These early syntheses were unsuccessful, with only macroscopic precipitates of polyaniline being obtained [9a,26]. Subsequently, this research team described the preparation of polyaniline colloids in aqueous media using functionalised, poly(ethylene oxide)-based stabilisers [26–29]. These authors believe that such stabilisers become chemically grafted to the polyaniline particles but have been unable to verify this hypothesis [46]. More recently, Liu and Yang have reported that colloidal polyaniline 'fibrils' can be prepared in the presence of apparently underivatised poly(acrylic acid) via an intermediate monomer/stabiliser gel phase [32].

In our own research programme we have focused on preparing and using tailor-made statistical copolymer stabilisers which contain pendant aniline groups [14,15,17–20] (see Fig. 3). In a typical colloidal polyaniline synthesis these stabilisers are 'aged' for 30–60 min. in aqueous acid solutions containing either the $(NH4)2S2O8$ or $KIO3$ oxidants prior to the addition of the aniline monomer (see Fig. 3). Our visible absorption spectroscopy studies have confirmed that the stabilisers' pendant aniline groups become oxidised under these conditions and so we believe that these activated groups inevitably participate in the subsequent in situ aniline polymerisation [15]. Thus the stabiliser becomes chemically grafted to the polyaniline particles. Certainly we have established that the pendant

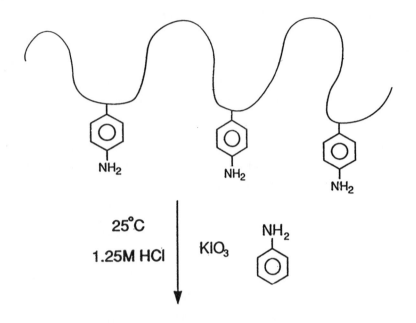

25°C

1.25M HCl   KIO₃

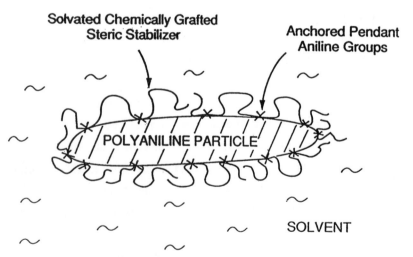

Fig. 3: General reaction scheme for the preparation of polyaniline colloids.

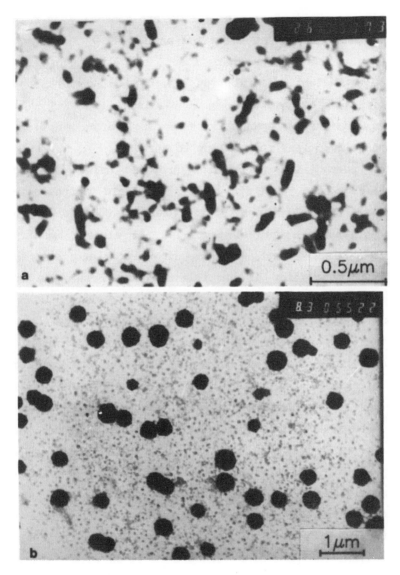

Fig. 4: Transmission electron micrographs ot polyaniline dispersions prepared with (a) KIO_3 and (b) $(NH_4)_2S_2O_8$ oxidants using a PNVP-AS stabiliser. Reproduced from Armes *et al.*, *J. Colloid Interface Sci.*, **118**, 410 (1987).

modification [15]. Such stabilisers are generally of rather broad molecular weight distribution ($M_w/M_n > 2$) but very recently we have shown that 'model' stabilisers of narrow molecular weight distribution ($M_w/M_n < 1.1$) based on poly(ethylene oxide) can also be utilised [19]. The mole % of pendant aniline groups in these polymeric stabilisers can range from less than 1 % up to 40 %.

The polyaniline dispersions prepared using the above stabilisers can exhibit various particle morphologies, ranging from highly anisotropic 'needles' to 'rice-grains' to polydisperse spherical particles. As already mentioned earlier, Vincent and Waterson have reported that this remarkable variation in particle morphology can, in some cases, depend on the stabiliser type [29]. However, we have recently shown that rather more subtle variations in the aniline polymerisation conditions (*e.g.* choice of chemical oxidant) can also significantly affect the particle morphology (see, for example, Figs 4a and 4b) [20].

We have determined the stabiliser/conducting polymer mass ratio for several of our polyaniline colloid systems using similar analytical methods to those described for the polypyrrole dispersions earlier (*e.g.* nitrogen microanalyses, F.T.I.R., visible absorption and [1]H n.m.r. spectroscopic assays of the post-reaction supernatant solutions *etc.* [14,15,17-20]). The stabiliser content of the polyaniline particles varies from 11-50 wt.% depending on the stabiliser type. Even for the highest stabiliser loadings these dispersions retain remarkably high electrical conductivity in the solid state.

(4) Other Systems:

In the last twelve months at Sussex we have developed two new colloidal forms of polyaniline. Firstly we have synthesised sterically-stabilised polyaniline particles directly in non-aqueous media *via* dispersion polymerisation [23]. More surprisingly, we have shown that stable polyaniline colloids can be easily prepared in the presence of small silica particles, *i.e.* using particulate rather than polymeric dispersants [37]. Such colloids comprise sub-micronic polyaniline-silica composite particles with a "raspberry" morphology (see Fig. 5); the polyaniline content of these composites can be easily varied between 20-50 wt.% depending on the synthesis conditions (*e.g.* type of chemical oxidant, silica particle size *etc.*). We have characterised these remarkable dispersions by a wide range of techniques. For example, CHN microanalyses and thermogravimetry were used to determine the stabiliser/polyaniline mass ratio [37]. Transmission electron microscopy, disc centrifuge photosedimentometry, photon correlation spectroscopy and charge-velocity analysis were each used to measure the particle size of one of these dispersions [38,39]. We found that this particular polyaniline-silica colloid had a broad, unimodal particle size distribution, with a mean weight-average particle diameter of 329 ± 71 nm. Other particle sizes can be prepared simply by changing the particle size of the original silica dispersant [47]. Finally, small-angle X-ray scattering has been utilised to probe the nanomorphology (internal structure) of the composite particles [38]. This experimental technique enabled us to estimate the average

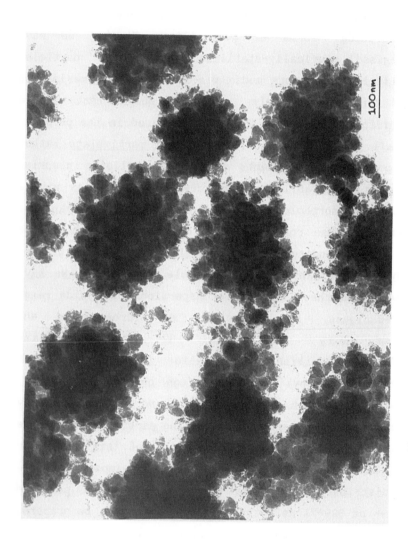

Fig. 5: Transmission electron micrograph of a polyaniline–silica composite colloid.

silica-silica separation distance within an individual polyaniline-silica "raspberry".

(5) Potential Applications:

It has become clear during our studies that, at least in the case of polypyrrole, these conducting polymer colloids are excellent "model" dispersions which possess both academic and commercial potential by virtue of a unique combination of physical properties. These include: remarkably high solid-state electrical conductivity; high intrinsic pigmentation; bio-compatibility; relatively high density (typically 1.50 g cm^{-3}); easily variable particle size (66-300 nm) and relatively narrow particle size distribution ($\pm$ 15 % standard deviation). In addition, and of no less importance, the synthesis of such dispersions can often be extremely facile in requiring only aqueous solutions, ambient temperatures, cheap reagents and commercially available polymeric stabilisers.

We are currently evaluating both these polypyrrole colloids and the polyaniline dispersions as anti-static additives for commodity plastics and commercial explosives at Sussex University. Other potential application areas cited in the literature for these systems include anti-static coatings for camera film [48], electrorheology [49], electrochromatography [46,50] and immunodiagnostics [51].

Conclusions

Colloidal dispersions of air-stable conducting polymers such as polypyrrole and polyaniline can be prepared *via* dispersion polymerisation utilising a wide range of polymeric and even particulate stabilisers. In general the stabiliser content of these conducting polymer colloids varies widely from 5-80 wt.% depending on the actual stabiliser/conducting polymer system. However, the solid-state conductivities of thin films or compressed pellets fabricated from such dispersions can be remarkably high (typically 0.1 to 1.0 S cm^{-1}), despite the presence of this electrically insulating stabiliser component. Thus these sterically-stabilised colloids represent a potentially useful and processable form of the normally intractable conducting polymer component.

A complete characterisation of these composite materials in terms of their particle morphology and stabiliser/conducting polymer mass ratio generally requires judicious selection (depending on the nature of the stabiliser) from a wide range of experimental techniques.

Acknowledgements

I wish to express my gratitude to the many staff members, students and technicians of the Universities of Bristol, Sussex and Salford and the Los Alamos National Laboratory, U.S.A. who have made invaluable contributions to this research programme. Financial support from the SERC, the Royal Society, the Nuffield Foundation, the Society for Chemical Industry, I.C.I. Resins, I.C.I. Agrochemicals, Courtaulds, Cabot Plastics and the D.R.A. is gratefully acknowledged.

References

1.   C. K. Chiang, C. R. Fincher, Y. W. Park, A. J. Heeger, H. Shirakawa, E. J. Louis, S. C. Gau and A. G. MacDiarmid, *Phys. Rev. Lett.*, 39, 1098 (1977).

2.   E. E. Havinga, L. W. van Morssen, W. ten Moeve, H. Wynberg and E. W. Meijer, *Polym. Bull.*, 18, 277 (1987).

3.   M. R. Bryce, A. Chissel, P. Kathirgamanathan, D. Parker and N. R. M. Smith, *J. Chem. Soc., Chem. Commun.*, 466 (1987).

4.   D. MacInnes and L. B. Funt, *Synth. Met.*, 25, 235 (1988).

5.   J. Yue and A. J. Epstein, *J. Am. Chem. Soc.*, 112, 2800 (1990).

6.   P. Kathirgamanathan, P. N. Adams, K. Quill and A. E. Underhill, *J. Mater. Chem.*, 1(1), 141 (1991).

7.   (a) A. Yassar, J. Roncali and F. Garnier, *Polym. Commun.*, 28, 103 (1987); (b) R. Partch, S. G. Gangolli, E. Matijevic, W. Cai and S. Arajs, *J. Colloid Interface Sci.*, 144(1) 27 (1991).

8.   R. B. Bjorklund and B. Liedberg, *J. Chem. Soc., Chem. Commun.*, 1293 (1986).

9.   (a) S. P. Armes, *Ph.D. thesis*, University of Bristol (1987); (b) S. P. Armes and B. Vincent, *J. Chem. Soc., Chem. Commun.*, 288 (1987).

10.   S. P. Armes, J. F. Miller and B. Vincent, *J. Colloid Interface Sci.*, 118(2), 410 (1987).

11.   S. P. Armes, M. Aldissi and S. F. Agnew, *Synth. Met.*, 28, 837 (1989).

12.   S. P. Armes and M. Aldissi, *Polymer*, 31, 569 (1990).

13.   S. P. Armes and M. Aldissi, *Synth. Met.*, 37, 137 (1990).

14.   S. P. Armes and M. Aldissi, *J. Chem. Soc., Chem. Commun.*, 88 (1989).

15.   S.P. Armes, M. Aldissi, S. F. Agnew and S. Gottesfeld, *Mol. Cryst. Liq. Cryst.*, 190, 63, (1990).

16.   S. P. Armes, M. Aldissi, G. C. Idzorek, P. W. Keaton, L. J. Rowton, G. L. Stradling, M. T. Collopy and D. B. McColl, *J. Coll. Interface Sci.*, 141, 119 (1991).

17.   S. P. Armes, M. Aldissi, S. F. Agnew and S. Gottesfeld, *Langmuir*, 6, 1745 (1990).

18.   R. F. C. Bay, S. P. Armes, C. J. Pickett and K. S. Ryder, *Polymer*, 32(13), 2456 (1991).

19.   P. Tadros, S. P. Armes and S. Y. Luk, *J. Mater. Chem.*, 2, 125 (1992).

20.   C. DeArmitt and S. P. Armes, *J. Colloid. Interface Sci.*, in the press.

21.   M. Beaman and S. P. Armes, accepted by *Coll. Polym. Sci.*

22.   Z. Rawi, J. Mykytiuk and S.P. Armes, submitted to *Colloids and Surfaces*.

23.   C. DeArmitt, S. P. Armes and D. Service, to be submitted to *J. Chem. Soc., Chem. Commun.*

24.   N. Cawdery, T. M. Obey and B. Vincent, *J. Chem. Soc., Chem. Commun.*, 1189 (1988).

25.   G. Markham, T. M. Obey and B. Vincent, *Colloids Surf.*, 51, 239 (1990).

26.   J. F. Miller, *B.Sc. thesis*, University of Bristol (1987).

27. E. C. Cooper, *Ph.D. thesis*, University of Bristol (1988).

28. E. C. Cooper and B. Vincent, *J. Phys. 'D'*, 22, 1580 (1989).

29. B. Vincent and J. Waterson, *J. Chem. Soc., Chem. Commun.*, 683 (1990).

30. F. Epron, F. Henry and O. Sagnes, *Makromol. Chem., Macromol. Symp.*, 35/36 527 (1990).

31. R. Odegard, T. A. Skotheim and H. S. Lee, *J. Electrochem. Soc.*, 138(10) 2930 (1991).

32. J.-M. Liu and S. C. Yang, *J. Chem. Soc., Chem. Commun.*, 1529 (1991).

33. E. Destryker and E. Hannecart, U.S. patent no. 5,066,706.

34. M. L. Digar, S. N. Bhattacharyya and B. M. Mandal, *J. Chem. Soc., Chem. Commun.*, 18 (1992).

35. P. Beadle, L. Rowan and S. P. Armes, manuscript in preparation.

36. K. E. J. Barrett, "Dispersion Polymerisation in Organic Media", Wiley, New York (1975).

37. M. Gill, J. Mykytiuk, S. P. Armes, J. L. Edwards, T. Yeates, P. J. Moreland and C. Mollett, *J. Chem. Soc., Chem. Commun.*, 108 (1992).

38. N. J. Terrill, T. Crowley, M. Gill and S. P. Armes, submitted to *J. Chem. Soc., Chem. Commun.*

39. M. Gill, S. P. Armes, D. Fairhurst, S. Emmett, T. Pigott and G. Idzorek, submitted to *Langmuir*.

40. S. P. Armes, M. Aldissi, M. Hawley, J. G. Beery and S. Gottesfeld, *Langmuir*, 7, 1447 (1991).

41.   G. Nimtz, A. Enders, P. Marquardt, R. Pelster and B. Wessling, *Synth. Met.*, <u>45</u>, 197 (1991).

42.   S. Y. Luk, C. DeArmitt, M. French, S. P. Armes and N. C. Billingham, manuscript in preparation.

43.   R. E. Myers, *J. Electr. Mater.*, <u>2</u>, 61 (1986).

44.   M. French, S. P. Armes and N. C. Billingham, manuscript in preparation.

45.   Y. E. Whang, J. H. Han, T. Motobe, T. Watanabe and S. Miyata, *Synth. Met.*, <u>45</u>, 151 (1991).

46.   B. Vincent, personal communication.

47.   F. L. Baines, M. Gill and S. P. Armes, manuscript in preparation.

48.   E. V. Thillo, G. Defieuw and W. De Winter, *Bull. Soc. Chim. Belg.*, <u>99</u>(11/12), 981 (1990).

49.   C. J. Gow and C. F. Zukoski, *J. Colloid Interface Sci.*, <u>136</u>(1), 175 (1990).

50.   H. Ge, P. R. Teasdale and G. G. Wallace, *J. Chromatography*, <u>544</u>, 305 (1991).

51.   P. J. Tarcha, D. Misun, M. Wong and J. J. Donovan, *A.C.S. PMSE Preprints*, <u>64</u>(1), 352 (1991).

Appendix

Stabiliser Abbreviations used in Tables 1-3

MC	Methyl(cellulose)
PNVP	Poly(N-vinylpyrrolidone)
PVA	Poly(vinyl alcohol-co-vinyl acetate)
PEO	Poly(ethylene oxide)
P4VP	Poly(4-vinylpyridine)
P4VP-BM	Poly(4-vinylpyridine-co-n-butyl methacrylate)
P2VP-BM	Poly(2-vinylpyridine-co-n-butyl methacrylate)
PVME	Poly(vinyl methyl ether)
PVAc	Poly(vinyl acetate)
PAA	Poly(acrylic acid)
PVA-d	Derivatised Poly(vinyl alcohol-co-vinyl acetate) containing pendant aniline groups
P2VP-AS	Poly(2-vinylpyridine-co-4-aminostyrene)
P4VP-AS	Poly(4-vinylpyridine-co-4-aminostyrene)
PNVP-AS	Poly(N-vinylpyrrolidone-co-4-aminostyrene)
PVIMZ-AS	Poly(N-vinylimidazole-co-4-aminostyrene)
PEO-An-PEO	Poly(ethylene oxide) of narrow molecular weight containing one aniline unit per PEO chain
PS-EA-PEO	Poly(styrene-co-ethyl acrylate-g-PEO)
PAC-g-PEO	'Poly(acrylate)' [29] with grafted PEO chains

COLLOID TITRATIONS AS A TOOL FOR OBTAINING INFORMATION ON THE PROPERTIES OF ADSORBED POLYAMINO ACIDS AND PROTEINS

J. Lyklema, W. Norde

Department of Physical and Colloid Chemistry
Agricultural University Wageningen
Dreijenplein 6
6703 HB Wageningen
The Netherlands

ABSTRACT

Conductometric and potentiometric titrations of polymer and other colloids in the presence of adsorbed or adsorbing proteins or polyelectrolytes are classical and powerful methods for obtaining information on the conformation and other properties of the macromolecules in the adsorbed state. This principle is illustrated by three examples: albumin adsorption on polystyrene latices and on silver iodide sols and polylysine on the same latex.

Potentiometric and conductometric acid-base titrations are traditional means of obtaining the surface charge density σ^0 of latices and other colloids. For easy reference we shall collectively call these techniques "colloid titrations". The technique allows one to determine σ^0 as a function of pH (for surfaces with variable charge) or the number of charged groups per unit area (for surfaces of constant charge). Similarly, polyelectrolytes and proteins in solution are routinely titrated to obtain the number of charges per molecule as a function of pH.

We now address the issue of titrations involving adsorbed or adsorbing polyelectrolytes and/or proteins, thereby posing the question of what useful information can be extracted on the mode of adsorption.

On protein structure changes and charge compensation

Protein adsorption is a complicated and fascinating phenomenon. One of the basic issues is the extent to which the conformation proteins have in the dissolved state persists upon adsorption. Experience has shown that in this respect a distinction should be made between structurally more stable, or "rigid" and structurally weaker, or "soft" proteins. The first category includes lysozyme, RNase, cytochrome-C and the subtilisins; the second class includes the albumines, caseines and the immunoglobins. Conformational changes may contribute to the affinity of a protein for a surface: groups originally buried inside a protein molecule may become exposed and bind to the adsorbate. Moreover, the entropy increase due to breakdown of secondary and tertiary structure also contributes to the driving force. However, structural changes can never be the sole driving force for protein adsorption, otherwise dissolved molecules would spontaneously change their conformations. Hence, there must be other interactions, which for the "rigid" proteins are the only ones. These include electrostatic, van der Waals, hydrogen-bond and hydrophobic attractions; in some cases specific covalent binding or ligand exchange may occur.

Electrostatic interactions may help the adsorption if protein and surface have opposite signs. Adverse electrostatic action may occur when these two have the same charge sign, but this effect is often not prohibitive: provided there is sufficient attraction of non-electrostatic origin, the electrostatic repulsion is readily blunted by ion uptake from the solution. Especially when all interactions tend to balance, ion incorporation can play a decisive role. Problems like these can be addressed by colloid titrations, because in that way information is obtained regarding the extent to which relevant groups remain free for titration after adsorption.

In this context it is interesting to compare surfaces of fixed charge (like polystyrene (PS) latex with sulfate groups) and those of variable charge (like silver iodide sols, where the surface charge σ^o depends on pAg = $-\log a_{Ag^+}$). For the fixed, charge adsorbents compensation can only be achieved by ion uptake from the solution, whereas for the latter system adjustment of the surface charge on the adsorbent is also possible. In terms of colloid titration, for a protein adsorbed on AgI two types of titration are viable: one to monitor the charge on the protein as a function of pH at given pAg, and the other to obtain σ^o for the adsorbent as a function of pAg at fixed pH, all of this at various amounts adsorbed, Γ, and indifferent electrolyte concentrations, c_s. On the other hand, for protein on sulfonated PS latex, only proton titration of the adsorbed protein is possible. The extent of charge compensation can only be inferred from other means, for instance by comparing the electrokinetic charge for (protein + latex) with that of protein and latex separately, or by measuring the co-adsorption of radiolabelled ions. Colloid titrations of proteins adsorbed on oxides are difficult to unravel since proton or hydroxyl ion uptake both stem from the protein and the adsorbent. In this paper we will not consider such systems.

In addition to these potentiometric colloid titrations we shall also present some results of conductometric titration in which a polypeptide is used as the titrant.

In passing, it may be added that another demonstration of the relevance of charge compensation is found in the fact that the pH where the plateau adsorption of "soft" proteins is a maximum coincides with the isoelectric point of the adsorbent-adsorbate complex: here electric contributions are maximally suppressed so that the protein retains its native conformation as much as possible.[1]

Experimental

In this paper we do not present much new material but reinterpret older data. Experimental details can be found in the relevant references. All experiments were performed under conditions where no measurable desorption upon pH-alteration took place. Additional adsorption during the titrations was avoided by replacing the equilibrium solution with a protein-free dialysate prior to titration.

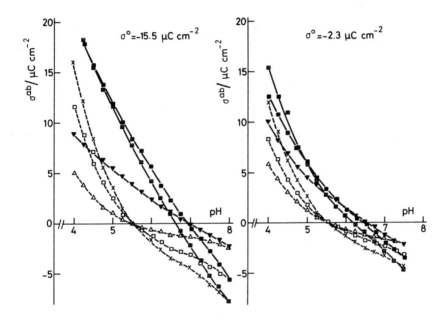

Figure 1. Acid-base titration charge on human serum albumin adsorbed on polystyrene latex with sulfate groups. The pH at which the protein was adsorbed was ■ 3.6; × 4.6; ▼ 7.3 or 8.0. c_{KNO_3}=0.05 M, temperature 25°C. Dashed curves (open symbols): hypothetical titration curves for HSA, side-on adsorbed on an uncharged surface.

Human serum albumin on PS latices

In figure 1 the titrated charge σ^{ab} of HSA is given, adsorbed on two different PS latices. It is expressed in μC per cm^2 of the adsorbent to facilitate comparison with the surface charge σ^o. The amount of adsorbed HSA changes with pH, passing through a maximum around pH 4.6. The dashed curves apply to the hypothetical case where HSA adsorbs side-on without undergoing any change in its σ^{ab}. The i.e.p. of HSA in 0,.05 M KNO_3 is 4.6. Hence, proteins adsorbed at pH 3 attached to the surface under electrostatically attractive conditions, those adsorbed at pH 4.6 under electroneutral and those at pH 7.3 or 8.0 under electrostatically repulsive conditions. From the figures it is concluded that in the range $\leq$ pH 7-8 upon adsorption the protein acquires a higher proton charge, which increases with increasing negative charge on the latex. This trend is interpreted as the result of proton uptake, or, for that matter, as a weakening of the acidity (or increasing basicity) of groups on the protein caused by the negative field of the latex. For pH < 6, which range we shall emphasize, mainly carboxyl groups are titrated. The weakening of these carboxyl groups is also observed when protein and latex have an overall attractive interaction. From this it is inferred that the carboxyl groups of the protein enrich the contact area.

Closer inspection of figure 1 also shows that the deviations from bulk solution behaviour are smallest at pH 4.6. This result is entirely in line with the general experience[2] that the structural alterations upon adsorption are relatively minor at the i.e.p. of the complex (which in this case coincides more or less with that of the protein). Adsorbing proteins are subject to more substantial conformational alterations at pH values away from the i.e.p. of the complex, probably because the molecule is then structurally more labile. In the titration curve, this effect results in a greater deviation from bulk behaviour for pH=3, and 7.3 or higher. However, as HSA adsorbs to PS latex at all pH, any electrostatic effect must be smaller than the non-electrostatic protein-latex interaction.

In figure 2 the same data are replotted as Henderson-Hasselbach (HH) curves for the pH range where carboxyl groups are titrated. Unlike figure 1, such a presentation only involves intensive quantities, so that comparison with the unadsorbed protein is straightforward.

For monofunctional polyelectrolytes the relation between pH, pK_a and the degree of dissociation can be written as

$$pH = pK_a - \log\left(\frac{1-\alpha}{\alpha}\right) + \frac{0.434 \Delta_{el}G(\alpha)}{RT} \qquad [1]$$

where $\Delta_{el}G(\alpha)$ is the Gibbs energy change caused by all electrical effects that make the polyelectrolyte different from the monomer. For free polyelectrolytes, $\Delta_{el}G(\alpha)$ includes the mutual interaction between the charges on the chain, the so-called polyelectrolyte effect. For an adsorbed polyelectrolyte, or for the carboxyl groups of an adsorbed protein in the appropriate pH-range, the contributions of the external field must be also included; these two contributions are not necessarily additive.

As the HH-plots appear to be linear over most of the range but with slopes n differing from unity it is advantageous to interpret them with Katchalsky's empirical equation[3]

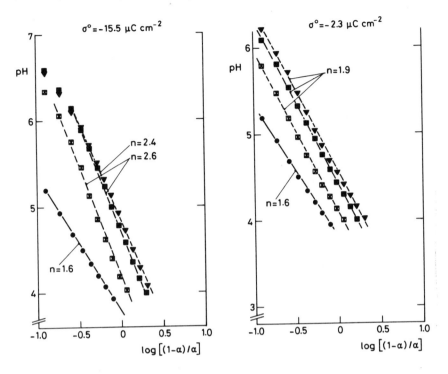

Figure 2. Henderson-Hasselbalch plot of the data of figure 1. The drawn curve with n = 1.6 refers to HSA in solution.

$$pH = pK_a' - n \log\left(\frac{1-\alpha}{\alpha}\right) \tag{2}$$

where n accounts for the polyelectrolyte effect. For free HSA the polyelectrolyte effect, as indicated by the index n, is 1.6, which is much higher than that for isolated aspartic and glutamic acid (n = 1.0). As expected, adsorption further increases n, the higher σ^o is. Here we are considering protein carboxyl groups in the contact region with the latex; those at the solution side remain virtually unaltered. In this contact region the dielectric constant is much lower than in water and if this were the only phenomenon, it would considerably increase electrostatic repulsion and hence $\Delta_{el}G(\alpha)$ or n. As in practice not so dramatic increments are observed, it is concluded that there must be substantial charge compensation in the system, most probably due to ion co-adsorption (see below). In terms of $\Delta_{el}G(\alpha)$, at given α the difference between the free and adsorbed protein is between 0.33 and 0.78 pH units, corresponding to increases of electric potential by not more than 40 mV.

It follows from the discussion so far that colloid titration is a powerful technique from which useful mechanistic information may be extracted. However, for the present system colloid titrations do not yield the amount of small ions (besides protons) that are co-adsorbed . For that, alternative techniques are required. In figure 3 an example is given, applying to the uptake of radiolabelled Na$^+$ ions[4]. Although the procedure is experimentally not easy, it is convincingly demonstrated that over the entire range the increase in charge density $\Delta\sigma$ due to Na$^+$ ions is positive with, as a trend, a minimum around the i.e.p. The value of $\Delta\sigma$ is not higher for the latex with the more negative charge, rather it reflects the number of protein carboxyls close to the PS surface; this number increases with increasing distance from the i.e.p. in both directions.

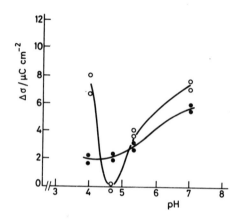

Figure 3. Co-adsorption of Na$^+$ ions for HSA adsorption on a PS latex. $\bullet$, σ^0 = -2.3 μC cm^{-2}; 0, σ^0 = -15.5 μC cm^{-2}.

Bovine serum albumin on silver iodide

The adsorption of BSA on AgI is similar to that of HSA on PS latices. The two proteins are almost identical and the two adsorbents are both predominantly hydrophobic, with a few charged sites. As a function of pH, the plateau adsorption again exhibits the typical curve with a maximum near the i.e.p.[5]. An advantage of the present system is that the charges on protein and adsorbent can be independently measured by colloid titration. (Under our experimental conditions the Ag$^+$ concentration is so low that it only adsorbs insignificantly on BSA, whereas protons are indifferent ions to AgI.) Alternatively, experiments can be done in a pH-stat or pAg-stat. In a pH-stat the AgI surface charge is titrated, whereby the amount of acid or base is measured, that is required to keep the pH constant. For the pAg-stat titration of the AgI-BSA complex a similar procedure is carried out, *mutatis mutandis.*

Figure 4 gives results from a pH-stat titration with KI and AgNO$_3$. In these titrations we face the problem that the specific surface area of suspended AgI, as measured by electrochemical methods (negative adsorption, double layer capacitance) is about three times higher than that via non-electrochemical procedures (dye-adsorption or BET)[6]. This old issue is not yet resolved. In the present case we have "mixed" conditions in that electrochemical and non-electrochemical adsorptions both contribute. The data of figure 4 refer to the non-electrochemical area; this brings the protein adsorption (in mg m^{-2}) in line with that on PS-latex and many other materials, but the charge in the absence of protein about threefold exceeds that of most AgI-interfacial electrochemistry[7]. For most of the following comparisons this choice has no consequences.

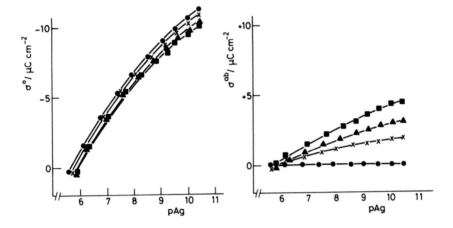

Figure 4. Colloid titration of AgI-sols in the presence of adsorbed BSA; amount adsorbed indicated. pH=5, electrolyte 0.1 M KNO$_3$. Temperature, 25°C. (a) Surface charge; (b) acid-base charge change of the protein. BSA adsorption (in mg m^{-2}) is 0 (●), 0.47 (×), 0.93 (▲) and 1.40 (■).

Figure 4a gives the change of the surface charge on the AgI, σ^o, due to increasing the amount of BSA adsorbed and figure 4b shows the accompanying proton uptake by the protein. For both the p.z.c. of pristine AgI surfaces (pAg=5.65) is chosen as the reference point. Comparison of the two diagrams shows that, under the prevailing conditions, in this "double-adjustable" system, the protein charge adjusts itself much stronger than σ^o does. A similar trend is observed at pH 4, where the protein is positive. However the trend is reversed at pH 6[5], where the protein carries a relatively large negative charge and consequently adsorbs in a thinner layer.

A comparison between the adsorption behaviour of BSA on AgI and HSA on PS-latices can be made by using figure 4a to identify the pAg for which σ^o is the same for the two systems. For example, at pAg=6,7 AgI carries a surface charge of -2.3 µC cm^{-2}, which is identical to σ^o of the low charged latex (figure 1b). According to figure 4b, at that pAg the protein takes up 1.3 µC cm^{-2} in proton charge as compared with adsorption on uncharged AgI. We can compare this with the difference at pH 5 between the curve for adsorption at pH 4.6 (protein uncharged) and that for a protein adsorbed at pH 5 (same charge on the protein as in the AgI experiment). The latter is not available but the difference must be close to that between pH's of adsorption of 4.6 and 4, which is about 1.5 μC cm^{-2}, i.e. comparable. Together with the similar amounts adsorbed on the two surfaces this suggests similarities in the behaviour of the protein upon adsorption on the two different adsorbents.

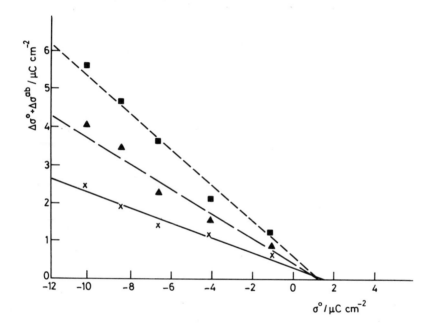

Figure 5. Sum adjustment of charge on the protein and on the surface as a function of the surface charge of the adsorbent prior to adsorption. pH=5. The amount of protein adsorbed (Γ in mg m^{-2}) is ■ 1.40 ▲ 0.93 and × 0.47.

Regarding the extent of charge compensation upon this process, with the AgI system more insight can be obtained than with the latex because the changes $\Delta\sigma^{o}$ and $\Delta\sigma^{ab}$ can be simultaneously measured. Figure 5 presents the sum $\Delta\sigma^{o} + \Delta\sigma^{ab}$ as a function of σ^{o}, the surface charge on the AgI prior to protein adsorption. It is seen that almost straight lines are observed, from which it may be inferred that substantial charge compensation takes place. As far as a small unbalance is observed, it is less than 2 μC cm^{-2}. This remaining deficit may be caused by co-adsorption of low molecular weight ions. In figure 3 we presented data on this co-adsorption for the latex system, these are very similar or somewhat larger. This difference, if real, may be caused by the fact that the latex system is mono-variant (only the protein charge can adjust) whereas the AgI -system is bi-variant (protein and surface charge can both adjust).

The main conclusion from these examples is that potentiometric colloid titrations are useful tools for obtaining interesting information on protein adsorption. As to the important role of conformational changes, it is underlined that any electrostatic antagonism is readily blunted by charge adjustment of the protein, and/or the surface (if it has a variable charge) and by incorporation of low molecular weight ions.

Polypeptides on PS latices

In general, polypeptides are much more flexible than proteins. Therefore, it must be expected that their conformation in the adsorbed state more closely obeys the rules for polyelectrolytes than those for proteins[8–10]. In the absence of added electrolyte, i.e. under conditions where the polyelectrolyte nature is most pronounced, the trend is that they adsorb as flat layers, loop formation being inhibited by the strong electrostatic repulsion that would otherwise arise. Being flexible, polyelectrolytes can bind in such a way that charge-charge attraction, if prevailing, is realized to a large extent without the need of co-adsorption of low molecular weight ions. For the interaction between a cationic and an anionic polyelectrolyte, where both components are flexible, this often leads to complexes with nearly or exactly 1:1 charge compensation[11]. However, when one of the components has charged groups in fixed positions (as is the case for sulfate groups on the surface of a latex) steric hindrance in the polycation way prevent such 1:1 binding. Here, colloid titrations again become useful. In particular, we shall consider conductometry for the system polylysine (PL) and polystyrene latex.

Acid-base conductometry of (dissolved) polyelectrolytes is a familiar technique for obtaining information on the polyelectrolyte conformation[12]. However, polyelectrolytes can also be used as the titrants. For the present case, titration of the latex with PL is very informative. Principles of the method have been described[13] and figure 6 is a typical illustration. In this figure, where the conductivity, κ, is plotted as a function of the added amount of PL, curve (1) represents the non-adsorbing situation: surface and polyelectrolyte are both positive; the slope of the straight line is a measure of the molar conductivity of the latter.

Curve (2) applies to the conditions where adsorption does occur. This curve consists of two linear parts separated by a short transition zone. The second branch has almost the same slope as in case (1), indicating that here titration of surface groups is completed: only the effect of added PL is recorded. In the strongly ascending first branch the conductivity increases with increasing amount of PL added in response to the binding of PL amino acid groups ($R'NH_3^+Br^-$) to protonated surface sulfate groups ($ROSO_3^-H^+$) Upon this binding H^+Br^- is liberated, causing a reduction in pH, see curve (3).

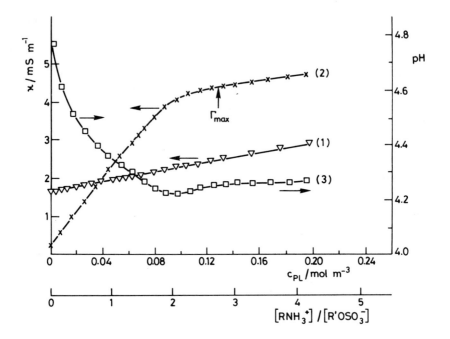

Figure 6. Conductometric colloid titration of positive PS latex (1) and negative PS latex (2) with polylysine, DP=1683. The pH is also given (3). No indifferent electrolyte added. Explanation in the text.

Regarding the conformation of the adsorbed polyelectrolyte, the following conclusions may be drawn.

(1) The ascending branch is linear up to slightly below the plateau value in the adsorption, indicated by Γ_{max} in the figure. This linearity indicates that over this range all molecules adsorb in the same conformation.

(2) If plotted as a function of ρ (the ratio of the numbers of $R'NH_3^+$ and $ROSO_3^-$ groups in the system) the linearity persists till about $\rho=2$, meaning that for each NH_3^+ group that binds one other NH_3^+ remains free (in the conductometric sense). This is in line with a very flat adsorption, though not so flat as to have all NH_3^+ groups on the surface. It is likely that such a maximally extended polyelectrolyte coil would be in an entropically very unfavourable conformation. This idea is supported by the observation that for the titration of aqueous solutions of sulfonated polystyrene with polylysine, the break point is found at $\rho=1$.

(3) From the linearity and the ratio of two it follows that between $\rho=0$ and $\rho=2$ the surface is covered by positive (where polylysine is adsorbed) and negative (uncovered) patches. Around $\rho=1$ the total surface charge is zero. Indeed, under these conditions the latices coagulate readily by "mosaic-type" charges[14] to become restabilized further on.

(4) After the point where $\rho=2$ the PL adsorption continues to increase somewhat. This means that other than electrostatic attractions must play a role. Probably hydrophobic bonding takes place between the lysine side chains and the essentially hydrophobic latex surface.

Conclusion

Our main conclusion is that relatively classical titration experiments can provide information on the conformation of adsorbed proteins and polyampholytes.

Acknowledgement

The authors appreciate the thorough way in which Dr. C. Haynes has read the manuscript.

References

1.   A.V. Elgersma, R.L.J. Zsom, W. Norde and J. Lyklema, Colloids Surf. 54 (1991) 89.
2.   W. Norde, Adv. Colloid Interface Sci. 25 (1986) 267.
3.   A. Katchalsky, P. Spitnik, J. Polym. Sci. 2 (1947) 1030.
4.   P. van Dulm, W. Norde and J. Lyklema, J. Colloid Interface Sci. 82 (1981) 77.
5.   J.G.E.M. Fraaije, W. Norde and J. Lyklema, Biophys. Chem. 41 (1991) 263.
6.   H.J. van den Hul, J. Lyklema, J. Am. Chem. Soc. 90 (1968) 3010.
7.   B.H. Bijsterbosch, J. Lyklema, Adv. Colloid Interface Sci. 9 (1978) 147.
8.   H.A. van der Schee, J. Lyklema, J. Phys. Chem. 88 (1984) 6661.
9.   J. Papenhuyzen, H.A. van der Schee and G.J. Fleer, J. Colloid Interface Sci,.104 (1985) 553.
10. M.R. Böhmer, O.A. Evers and J.M.H.M. Scheutjens, Macromolecules 23 (1990) 2288.
11. B. Phillipp, W. Dawydoff and K.-J. Linow, Z. Chem. 22 (1982) 1.
12. H. van Leeuwen, R.F.M.J. Cleven and P. Valenta , Pure Appl. Chem. 63 (1991) 1251; M. Mandel in *Encyclopedia of Polymer Sci. and Engineering.* H. Mark et al. Eds., Vol. 11, 2nd ed. John Wiley (1988), especially pages 811-19.
13. B.H. Bonekamp, J. Lyklema, J. Colloid Interface Sci. 113 (1986) 67.
14. J. Gregory, J. Colloid Interface Sci. 42 (1973) 443.

BRIDGING AND DEPLETION FLOCCULATION OF LATEX PARTICLES BY ADDED POLYELECTROLYTES

Nick Cawdery and Brian Vincent*
Physical Chemistry, University of Bristol, Cantock's Close,
Bristol BS8 1TS, U.K.

ABSTRACT

Bridging or depletion flocculation may occur in particulate dispersions when polymer is added. In this work, both effects have been observed in the following system: aqueous dispersions of polystyrene (PS) latex particles, carrying terminally-anchored poly(ethylene oxide) (PEO) chains, to which poly(acrylic acid) (PAA) has been added. The poly(acrylic acid) was prepared in monodisperse form by the hydrolysis of poly(t-butylacrylate), prepared by an anionic polymerisation route.

In solution, PEO is known to coacervate with PAA at low pH ($< \sim 3.5$), (through H-bonding), whereas at higher pH's ionisation of the carboxylic acid groups on the PAA, prevents any interaction. Thus, by analogy, the PS-g-PEO particles were expected to adsorb PAA at low pH, but not at high. This is what was found. Moreover, bridging flocculation was observed at low pH values, and depletion flocculation at higher pH values, each over certain concentrations of PAA, depending on the background electrolyte concentration. A semi-quantitative interpretation of the depletion results is presented.

INTRODUCTION

It is well known (Napper, 1983) that when polymers,
including polyelectrolytes, are added to a colloidal
dispersion the stability of that dispersion may be enhanced
or reduced. In the case of added homopolymers, in a good
solvent environment, there are basically two conditions,
under which attractive interparticle interactions, and hence
instability, may be induced: (a) for adsorbing polymers,
at low surface coverages, bridging flocculation may occur,
as a result of the co-adsorption of polymer chains originally
on one particle onto a second particle; (b) for non-adsorbing polymers, the creation of a polymer depletion zone near
the particle surface produces an osmotic attraction between
two such particles when their depletion zones overlap,
giving rise to depletion attraction.

Whether a polymer adsorbs onto a surface or not depends
on the χ_s parameter i.e. the net surface/segment-surface/
solvent interaction parameter. There exists a critical value
for χ_s (χ_{sc}) only above which adsorption occurs (Cohen
Stuart et al., 1986). It is possible to vary χ_s systematically by varying the solution conditions. In this study we
do this for the following system: aqueous dispersions of
polystyrene (PS) latex particles, carrying terminally-
anchored poly(ethylene oxide) (PEO) chains (i.e. PS-g-PEO
particles), to which poly(acrylic acid) (PAA) is added.
PAA has a pK_a value of 5.5 (Kawaguchi and Nagasawa,
1969). Hence by varying the solution pH, the degree of
ionisation of the acid hydrogen on the PAA chains may be
varied. At low pH values, there is little ionisation, and

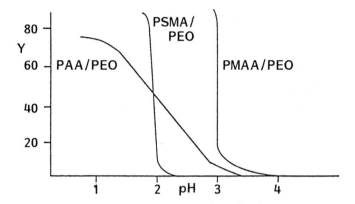

Fig. 1: pH dependent of precipitate yield (Y%) for various poly(carboxylic acids) + PEO. PAA= poly(acrylic acid); PMAA = poly(methacrylic acid); PSMA = a copolymer of styrene and maleic acid (Ikawa et al., 1975).

the PAA chains can be expected to "adsorb" onto the PEO chains of the latex particles (through H-bonding), whereas at high pH values no such adsorption is expected. This situation reflects the well-established behaviour of mixtures of aqueous PEO and PAA solutions: coacervation is observed at low pH values, but not at high pH values (Bailey, 1964). Figure 1 shows the pH dependence of the precipitate yield for mixtures of various poly(carboxylic acids) and PEO (Ikawa et al., 1975).

For the PS-g-PEO latex plus PAA systems, therefore, bridging flocculation may be anticipated at <u>low</u> pH values. Indeed, Evans and Napper (1973) have already reported such behaviour. Moreover, a similar recipe may be used in practice in "electrocoat" paint systems (Doroszkowski, A., 1986). However, as far as the authors are aware, no

studies have been reported of the possible depletion flocculation behaviour which may be expected at high pH values.

In this paper, therefore, we report studies on this aspect,
and compare the high pH and low pH adsorption and stability behaviour of PS-g-PEO dispersions, in the presence of
added PAA, as a function of the pH and ionic strength of
the aqueous medium.

It should be noted that, in comparison to work with neutral
polymers, there have been relatively few studies of
depletion flocculation with polyelectrolytes. Snowden et al.
(1991) studied the effects of sodium carboxymethylcellulose
and sodium poly(styrene sulphonate) on the stability of
aqueous silica dispersions, whilst MacMillan (1989), in this
laboratory, made similar studies with poly(acrylic acid).
Rawson et al. (1988) also in this laboratory, had previously
investigated the effect of sodium poly(styrene sulphate) on
similar aqueous dispersions of PS-g-PEO particles to those
discussed in this paper. This was the first reported case
of polyelectrolyte depletion we are aware of.

EXPERIMENTAL

Polystyrene-g-Poly(ethylene oxide) (PS-g-PEO) Particles

A PS-g-PEO latex was prepared using the method described
by Cowell and Vincent (1982) and by Bromley (1985). This
involves the dispersion copolymerisation of styrene and PEO-
methacrylate together with a small quantity of divinyl-
benzene as cross-linker, using α-azoisobutyronitrile as the
initiator, in aqueous media. The mean-particle diameter of

the latex particles was found from transmission electron microscopy to be 460 ± 37 nm. The PEO chains had an $\overline{M}_n$ value of 2000. Cosgrove and Ryan (1990) have reported a value of 3.6 nm for the thickness (δ) of the hydrated PEO sheath on similar latex particles. The critical flocculation temperature of the PS-g-PEO latex in 0.26 mol dm^{-3} aqueous $MgSO_4$ solution was found to be 58 + 1°C. This value is identical to the value reported previously by Bromley (1985) for a similar latex, and is close to the theta-temperature for PEO in 0.26 mol dm^{-3} $MgSO_4$ solution (Boucher and Hines, 1976).

In order to check for any possible electrostatic interactions between the latex particles, their electrophoretic mobility was monitored as a function of pH and NaCl concentration,

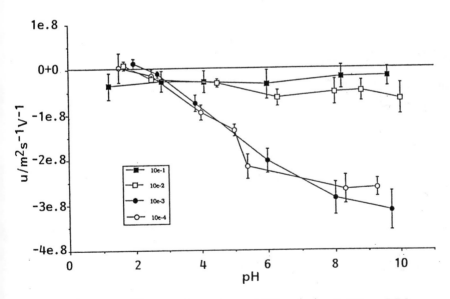

Fig. 2: Electrophoretic mobility (u) of PS-g-PEO particles as a function of pH and NaCl concentration.

using a "Pen Kem 3000" apparatus. The results are shown
in figure 2. As may be seen, the particles become increasingly negatively charged with increasing pH, particularly
at low ionic strengths; this effect was found to be
reversible. The maximum mobility value at pH 9 (-3 x
$10^{-8} m^2 s^{-1} V^{-1}$) in 10^{-4} mol dm^{-3} NaCl solution, corresponds
to a zeta potential of -46 mV. The charge is probably due
to the presence of weak carboxylic acid groups at the polystyrene core surface, arising from the initiator.

Poly(acrylic acid) PAA

I) Preparation:

Samples of PAA having a low dispersity in molecular mass
were obtained by the acid hydrolysis of poly(t-butyl
acrylate) (PTBA) which can be prepared by anionic polymerisation reactions (Miller and Raubut, 1958) using
initiators which involve Li^+ counter ions (Fayt et al.) The
polymerisation reactions were carried out on a custom-
built vacuum line apparatus, fitted with an oil-diffusion
pump to obtain pressures < 10^{-5} torr. All glassware used
was vigorously cleaned and dried.

The preparation route used for the synthesis of PTBA is
summarised in figure 3. The diphenylethylene (DPE) t-
butyl lithium adduct (I)[*] was prepared first in order to
provide an end-group functionality on the PTBA (and
subsequent PAA) chains, which could be utilised as a uv
chromophore for determining polymer concentrations.

[*] In some cases α-methylstyrene (α-MS) was used
instead.

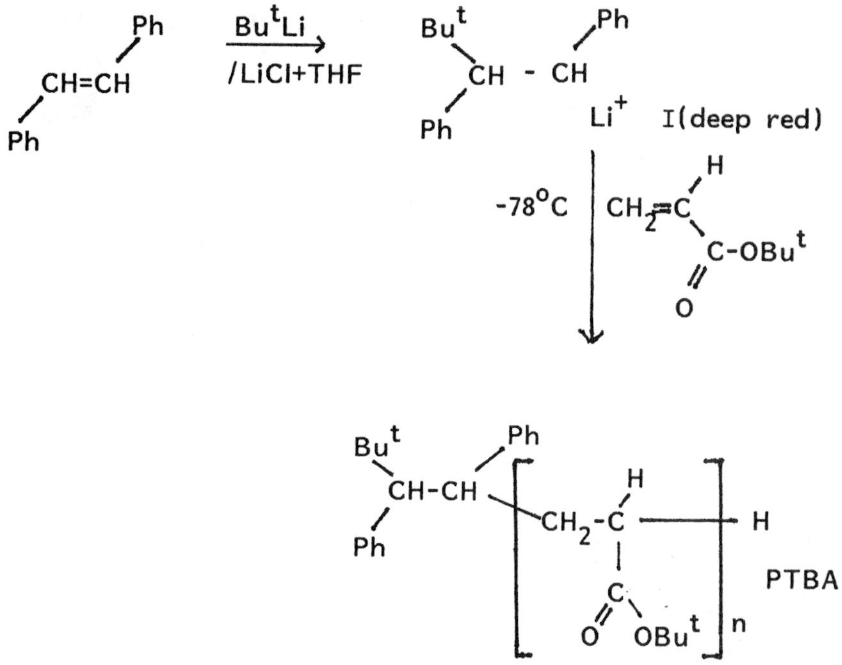

Fig. 3: Preparation of PTBA by anionic polymerisation.

The solvent used was THF. This was initially stored over CaH_2 and then distilled onto fresh CaH_2. It was then degassed by several freeze/thaw cycles and then distilled onto a sodium/naphthalene mixture, prior to distillation directly into the reaction vessel. LiCl was placed in the reaction vessel prior to adding the THF. The amount of LiCl was fixed in relation to the amount of t-butylacrylate monomer to be subsequently polymerised. The ratio LiCl: monomer is recommended to be $\sim$ 10:1 for PTBA molecular masses $< 10^5$ (Fayt et al., 1987) and $\sim$ 25:1 for PTBA molecular masses $> 10^5$ (Klein et al., 1990). The role of the extra Li^+ counter ions is to moderate the very fast polymerisation kinetics of t-butylacrylate in THF, even at $-78^\circ C$.

Diphenylethylene was added to the LiCl/THF mixture using a degassing syringe, through a subaseal, and this solution was accurately titrated with t-BuLi until a persistent red colour was obtained, indicating any impurities which might react with t-BuLi had been eliminated. The required amount of t-BuLi to form the 1:1 adduct (I) was then added; this had a deep red colouration.

t-Butylacrylate monomer was stored at $4^{\circ}C$ over CaH_2, and distilled onto fresh CaH_2 prior to use. It was then titrated with a 25% solution of triethylaluminium in hexane, until a pale yellow colour persisted. After 10 min the t-butyacrylate monomer was distilled into a volumetric column above the reaction vessel, prior to addition to the reaction vessel containing the solution of the adduct I in the THF/LiCl mixture. The polymerisation reaction was allowed to proceed for $\sim$ 2h at $-78^{\circ}C$, and then terminated by the addition of methanol.

The PTBA formed was purified by precipitation from methanol, freeze-dried from 1,4-dioxan and finally dried at $80^{\circ}C$.

Hydrolysis of PTBA to PAA was effected as follows (Miller et al., 1960). A solution of PTBA in 1,4-dioxan was refluxed for 1h and then HCl was added and the solution further refluxed for 24h. The resulting PAA solution was exhaustively dialysed against distilled water, and then freeze-dried from water and finally dried at $80^{\circ}C$.

II) Characterisation:
Because GPC analysis of aqueous solutions of polyelectrolytes is not straightforward the GPC analyses were carried

Table I: GPC data for the PTBA polymers hydrolysed to PAA

CODE	REFRACTIVE INDEX DETECTOR		VISCOMETRY DETECTOR		M_n for PAA
	M_n	M_w/M_n	M_n	M_w/M_n	
PTBA 10	16700	1.24	19800	1.01	11000
PTBA 11	76300	1.23	86700	1.09	48000
PTBA 14	170700	1.18	197300	1.13	111000
PTBA 15	105700	1.20	119400	1.09	67000

out[*] on the parent PTBA samples in THF, using polystyrene standards for calibration; this is thought to be a good approximation in determining molecular masses for PTBA (Klein et al., 1990). Two detectors were used: a refractive index detector and a viscometry detector. The results are compared in table I, where it can be seen that the agreement is reasonable between the two sets of results. The four PTBA samples listed, which are all reasonably monodispersed were the ones which were subsequently hydrolysed to PAA. In the last column the corresponding M_n value for PAA is given (56.3% of the corresponding M_n value for the parent PTBA).

^{1}H NMR analysis of the PAA samples showed that complete hydrolysis of the parent PTBA samples had been achieved. This was monitored by noting the disappearance of the distinctive t-butyl peak at 1.2 ppm.

Solutions of the various PAA samples were also conductiometrically titrated with sodium hydroxide. In all the cases the amount of sodium hydroxide required to reach the neutralisation point in the titration, implied that more than 98% of the t-butylacrylate groups had been hydrolysed to the carboxylic acid functionality, supporting the NMR results referred to above.

Adsorption Isotherms
Adsorption isotherms for the PAA samples onto the PS-g-PEO particles, at a fixed background electrolyte (NaCl)

* Thanks to Mrs L.A. Catt at I.C.I. C&P plc for running these samples.

concentration and pH, were determined by tumbling mixtures for 24h, centrifuging the latex particles and then quantitatively analysing for the equilibrium PAA concentration (c_z) is the supernatant. To this end use was made of the u.v. absorption peak at a wavelength of 215 nm for the diphenyl chromophore. Knowing also the initial PAA concentration, the adsorbed amount (Γ/mg m^{-2}) may be readily calculated.

Stability Studies

The following solutions were prepared at the required pH values:-

1) PS-g-PEO latex of known particle volume fraction and at twice the required final NaCl concentration.
2) PAA solution
3) Water

x g of (2) was then mixed with y g of (1) (x+y=1) in a small sample tube, before adding 1 g of (1). The final PAA concentrations are therefore quoted as weight fractions (W_2) (rather than volume fractions, ϕ_2). In a given experiment a series of stoppered tubes was prepared containing mixtures for which all parameters were kept constant, except that the PAA concentration was systematically increased.

The tubes were allowed to stand at ambient temperature for 24h and evidence for flocculation assessed visually. This was quite straightforward and reproducible, for latex volume fractions in the range, $\phi = 5 \times 10^{-4}$ to 10^{-1}. In some cases turbidity measurements were also made. A distinct change in turbidity occured with increasing PAA concentration, corresponding to the onset of flocculation/ restabilisation. Good agreement with the direct visual assessment method was obtained. The PAA weight fraction where flocculation is first observed and that where re-stabilisation is observed, are designate $W_2^\dagger W_2^\ddagger$, respectively.

Osmotic Pressure Measurements

The osmotic pressure of the PAA solutions, at a given pH and electrolyte concentration, were determined using a Knauer membrane osmometer with a cellulose acetate membrane. All measurements were carried out at 25.0 ± 0.1°C.

RESULTS

Adsorption Isotherms

Adsorption isotherms for PAA11 (table I) onto the PS-g-PEO latex particles, at 10^{-2} mol dm^{-3} NaCl solution and various pH values in the range 2-4, are shown in fig. 4. As may be seen, the adsorbed amount descreases, at fixed PAA concentration, with increasing pH, until at pH 4.0 very little adsorption occurs. These results reflect the PAA/PEO coacervation behaviour shown in fig. 1. Clearly, at pH > 4 only a small fraction of the carboxylic acid groups of the PAA chains remain undissociated, so that H-bonding with the PEO chains on the latex particles (i.e. χ_s) is greatly reduced.

With increasing NaCl concentration, it was found that the PAA adsorbed amount, at a given pH value, increased slightly ($\sim$ 20% over the NaCl concentration range 10^{-4} to

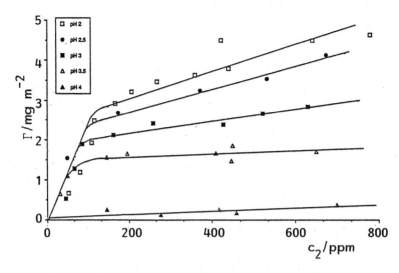

Fig. 4: Adsorbed amount (Γ) as a function of equilibrium PAA concentration (c_2), at various pH values. [NaCl] = 10^{-2} mol dm^{-3}.

10^{-1} mol dm^{-3}). This presumably reflects the compaction of the chains as the intra-chain repulsion is increasingly screened with increasing NaCl concentration. Virtually no dependence of the adsorbed amount on PAA molecular weight was found, for the four polymers listed in table I.

Stability Studies

Fig 5 shows a "stability map" for the PS-g-PEO latex plus PAA 14 (table I) in a background NaCl concentration of 10^{-4} mol dm^{-3}. This takes the form of a PAA wt. fraction (W_2) versus pH graph. Two instability regions may be identified: one below pH 2.8, the other above pH 3.5. In the low pH region, where PAA adsorption occurs, "fluffy" flocs form very quickly and seemingly irreversibly, over a certain range of polymer wt. fractions ($W_2^+ < W_2 < W_2^{\ddagger}$). This

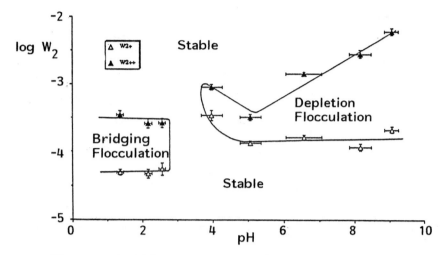

Fig. 5: Stability map for PS-g-PEO latex plus PAA 14
$\phi = 5 \times 10^{-3}$;   [NaCl] = 10^{-4} mol dm^{-3}.

would seem to be bridging flocculation. In the high pH
region, on the other hand, where (virtually) no PAA
adsorption occurs, much weaker flocculation occurs, again
over a certain PAA wt. fraction range. In this case
flocculation occurs much more slowly ($\sim$ 24h), and the
system effectively forms two co-existing colloidal "phases":
a low volume fraction ("gas-like") phase in equilibrium
with a condensed ("solid-like") phase. This is typical of
depletion flocculation behaviour, as described previously,
for example by Vincent et al. (1986). No flocculation was
observed over the pH range 2.8 to 3.5.

Fig. 6 shows a similar stability map to that shown in fig. 5,
except that the background NaCl concentration has been
raised to 10^{-1} mol dm^{-3}. Again, two instability regions
occur, one at low pH, the other at higher pH values.

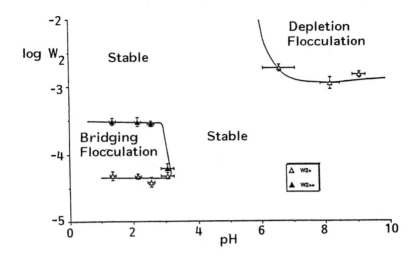

Fig. 6: Stability map for PS-g-PEO latex plus PAA 14 $\phi = 5 \times 10^{-3}$; [NaCl] $= 10^{-1}$ mol dm^{-3}.

There seems to be very little effect of salt concentration on the bridging flocculation region at low pH, but there is a much stronger effect on the depletion flocculation region at high pH. The onset of depletion flocculation occurs at a higher minimum pH value: ~ 6 at 10^{-1} mol dm^{-3} NaCl. A much higher value of $W_2^\dagger$ value is necessary at the higher NaCl concentration. Also no restabilisation could be achieved by the highest PAA concentration ($W_2 \sim 0.05$) which could be studied before the high viscosity of the solution made the experiments intractible. The general effect of increasing NaCl concentration on $W_2^\dagger$ is shown in fig. 7

Also shown in fig. 7 is the effect of increasing MgCl$_2$ concentration on $W_2^\dagger$. At a given molar electrolyte concen-

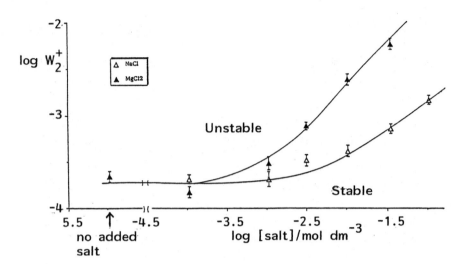

Fig. 7· Effect of [NaCl] and [MgCl$_2$] concentration on $W_2^\dagger$. ϕ = 5 x 10^{-3}.

tration it is seen that $W_2^\dagger$ is greater in the case of MgCl$_2$ than for NaCl, i.e. the dispersion is _more_ stable to the addition of PAA in the presence of MgCl$_2$, compared to NaCl.

In all the experiments described so far the volume fraction (ϕ) of the latex particles was fixed (at 5 x 10^{-3}). The effect of varying ϕ, over the range from 5 x 10^{-4} to 10^{-1}, on $W_2^\dagger$ and $W_2^\ddagger$ was investigated for both the bridging and depletion flocculation regions (at 10^{-4} mol dm^{-3} NaCl). In the case of bridging flocculation it was found that both $W_2^\dagger$ and $W_2^\ddagger$ increased steadily with increasing ϕ. In the case of depletion flocculation, however, both $W_2^\dagger$ and $W_2^\ddagger$ _decreased_ with increasing ϕ. The effect of varying the molecular weight of the PAA used (over the range indicated in table I) was also investigated. In the case of bridging flocculation, at low pH values, little effect of molecular weight variation was observed for $W_2^\dagger$, although $W_2^\ddagger$ decreased slightly with increasing molecular weight of PAA. In the case of depletion flocculation, at higher pH values, flocculation was only observed with PAA samples 14 and 15 (M.W. 111,000 and 67,000, respectively). No flocculation was seen with the lower M.W. PAA samples (table I), over the NaCl concentration range 10^{-4} to 10^{-1} mol dm^{-3}. A "cut-off", minimum molecular weight of added polymer, for the onset of depletion flocculation, has been reported for other systems. A recent example is for dispersions of silica particles, carrying terminally-anchored polymethyl-methacrylate chains, in dioxan, to which free polymethylmethacrylate was added (Milling et al., 1991).

DISCUSSION

I) Bridging Flocculation

It is clear that the flocculation behaviour observed at low pH's is bridging flocculation, and reflects the adsorption of PAA onto the PS-g-PEO particles at coverages less than saturation (fig. 4). The $W_2^\dagger$ and $W_2^\ddagger$ "boundaries" shown in the stability maps (figs. 5 and 6) are not strictly "thermodynamic" boundaries in this case; rather, they are "kinetic" boundaries. They define the range of W_2 values where the rate of flocculation is sufficiently fast that flocs become visible over a timescale of minutes.

Since the PS-g-PEO particles have only small electrophoretic mobilities in this pH range (fig. 2), corresponding to a maximum value for the zeta potential of ~5 mV, electro-static repulsion between the particles can be ignored. It is not too surprising, therefore, that little effect of NaCl concentration on $W_2^\dagger$ and $W_2^\ddagger$ is observed for the bridging flocculation regime.

The fact that both $W_2^\dagger$ and $W_2^\ddagger$ increase with increasing particle volume fraction, ϕ, simply reflects the increasing particle surface area, and therefore the higher initial PAA concentration required to achieve a similar level of coverage.

No quantitative analysis of the bridging flocculation results is attempted here.

II) Depletion Flocculation

As discussed in the "Results" section, in the high pH flocculation regimes (figs. 5 and 6), where weak depletion flocculation is occurring the systems effectively phase separate into two co-existing colloidal phases. This is a phenomenon generally observed in the case of depletion flocculation (Emmett and Vincent, 1990). Further evidence for this is the fact that $W_2^\dagger$ and $W_2^\ddagger$, in this case, decrease with increasing particle volume fraction (i.e. the opposite trend to that observed in the case of bridging flocculation). This behaviour has also been observed before, for example, by Cowell and Vincent (1978), who studied the effect of adding the neutral polymer, poly-(ethylene oxide), also to PS-g-PEO particles, similar to those used in this work. These authors interpreted their results in terms of eqn. (1):

$$\Delta F_f = \Delta U_f - T\Delta S_f \qquad \ldots \ (1)$$

ΔF_f, ΔU_f and ΔS_f are the free energy, energy and entropy change associated with the flocculation process. The flocculation boundaries ($W_2^\dagger$ and $W_2^\ddagger$) are now true thermodynamic boundaries given by the condition $\Delta F_f = 0$. Outside the flocculation region ΔF_f is positive and the dispersions are thermodynamically stable. In eqn. (1) both ΔU_f and ΔS_f are negative quantities. ΔU_f is essentially a function of the depth of the (shallow) energy minimum (V_{min}) into which particles flocculate. ΔS_f, on the other hand, is essentially a function of ϕ; the higher ϕ the lower $|\Delta S_f|$. Thus, as ϕ increases, the value of $|V_{min}|$ required to induce flocculation is reduced. Thus, less added free polymer is required (Cowell et al., 1978) i.e. $W_2^\dagger$ is reduced.

In an attempt to calculate trends in V_{min} values for the systems studied here, a pragmatic model was used, following that proposed by Fleer, Scheutjens and Vincent (SFV, 1984). These authors showed that, for dispersions of hard sphere particles, in the presence of free polymer, the depletion energy (V_{dep}) as a function of particle separation (h) is given by,

$$V_{dep} = 2 \pi a \Pi \Delta^2 \left(1 - \frac{h}{2\Delta} \right) \qquad (\Delta \ll a) \qquad \ldots (2)$$

or when h = 0 (particle contact),

$$V_{dep} = 2 \pi a \Pi \Delta^2 \qquad \ldots (3)$$

where a is the particle radius, Π is the osmotic pressure of the bulk polymer concentration, and Δ is the effective depletion layer thickness for polymer chains at a non-adsorbing surface ($\chi^s < \chi_c^s$). At low polymer concentration

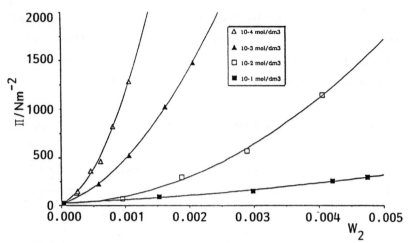

Fig. 8: Osmotic Pressure (Π) as a function of wt.
function (W_2) of PAA in various NaCl solutions. PAA 14
at pH 9.

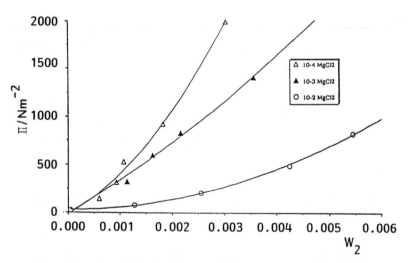

Fig. 9: Osmotic Pressure (Π) as a function of wt.
fraction (W_2) of PAA in various $MgCl_2$ solutions. PAA 14
at pH 9.

the SFV theory predicts that $\Delta = 1.4 \; r_g$, where r_g is the radius of gyration of the polymer coils in dilute solution

Thus evaluation of $V_{dep}(h)$, as a function of bulk PAA weight fraction (W_2) or volume fraction (ϕ_2), requires knowledge of $\Pi(\phi_2)$ and $\Delta(\phi_2)$. The former quantity can be obtained directly from osmotic pressure measurements. Figurs 8 and 9 show the $\Pi(W_2)$ data obtained in this work for PAA solutions, at different NaCl and $MgCl_2$ concentration, respectively. It can be seen that, at a given W_2 value and molar electrolyte concentration, Π is greater in the case of NaCl than $MgCl_2$. This is due to the greater screening effect of Mg^{2+} ions, comapred to Na^+ ions, on the intra-chain repulsion for the polyelectrolyte.

According to eqn. (3), an increase in Π would, in itself lead to an increase in V_{dep} and hence a smaller value for $W_2^{\dagger}$. This argument would account <u>broadly</u> for the following observations (fig. 7):

(i)   for a given electrolyte type: the value of $W_2^{\dagger}$ increases with increasing electrolyte concentration.

(ii) at a given electrolyte concentration: $W_2^{\dagger}$ is greater for $MgCl_2$ than for NaCl.

However, in order to obtain a more <u>quantitative</u> interpretation of the experimental observations, the following points must also be considered:

(a)   account must also be taken of any variation in $\Delta(W_2)$ in calculating $V_{dep}(h)$, using eqn. (2).

(b)   the total interparticle interaction energy $V(h)$ is given by

$$V(h) = V_A(h) + V_E(h) + V_{dep}(h) \qquad \dots (4)$$

where $V_A(h)$ is the van der Waals attraction between the particles and $V_E(h)$ is the electrostatic (electrical double layer) interaction energy, associated with the particle charge at high pH (fig. 2).

With regard to (b) above, it may be shown (Cawdery, 1992) that $V_A(h)$ is negligible if the assumption is made that the Hamaker constant of the (hydrated) PEO sheath around the PS particle cores is the same as that of the medium. $V_E(h)$ may be calculated using the following standard expression,

$$V_E = 2 \pi \varepsilon a \; \zeta^2 \; \ln[1 + \exp(-\kappa h)] \qquad \dots (5)$$

In eqn. (5), ε is the permittivity of the medium; ζ is the zeta potential of the particles, calculated from the experimental electrophoretic mobility values, using the O'Brien and White (1978) formulation. In calculating ζ and κ (the Debye-Hückel screening parameter), consideration has to be given to the value of the electrolyte concentration between the particles. When two particles approach to a separation less than 2Δ, then PAA chains are effectively excluded from the region between the particles. Hence, a Donnan salt effect is set up, such that the electrolyte concentration between the particles is greater than in the bulk solution containing the polyelectrolyte chains. The standard correction term for the Donnan effect across a semi-permeable membrane was applied in calculating the electrolyte concentration between the particles ($h < 2\Delta$).

Taking account of effect (a) above is much more problematical. There is no analytical theory as yet for $\Delta(\phi_2)$

for polyelectrolytes. Vincent (1990) has derived the following analytical expression for $\Delta(\phi_2)$ for <u>neutral</u> random coil polymers,

$$\frac{\Delta}{\Delta_o} - \frac{1}{\Delta} = -\frac{\Pi}{2kT}\left(\frac{v_2}{\phi_2}\right)^{2/2} \qquad \ldots (6)$$

where Δ_o is the value of Δ in the limit of zero polymer concentration and v_2 is the molar volume of the polymer chains. However, polyelectrolyte chains only approach random coil behaviour at <u>high</u> background electrolyte concentrations. At <u>low</u> background electrolyte concentration the behaviour is more rod-like. For the case of actual rigid rods, Auvray (1981) had earlier suggested that Δ is <u>independent</u> of ϕ_2.

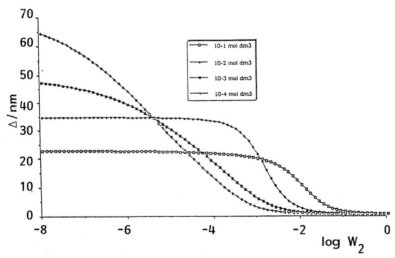

Fig. 10: Apparent depletion layer thickness (Δ) as a function of PAA wt. fraction (W_2) for PAA 14, calculated using the theory of Vincent (1990) for various NaCl concentrations.

Figure 10 shows the dependence of Δ on W_2 for PAA 14, calculated at various electrolyte concentrations according to the theory of Vincent (1990), i.e. assuming random-coil behaviour. The value for Δ_o was taken to be 1.4 r_g following the SFV theory (see earlier). Values for r_g were taken from the literature for sodium polyacrylate in NaCl (Takahashi and Nagasawa, 1964).

Based on these assumptions, V(h) curves were calculated using eqn. (4). Results for PAA 14 at pH 9 are shown in figures 11 and 12, for 10^{-1} mol dm^{-3} and 10^{-2} mol dm^{-3} NaCl, respectively. In both cases four W_2 values were selected, scanning the range between the experimentally observed $W_2^{\dagger}$ and $W_2^{\ddagger}$ values.

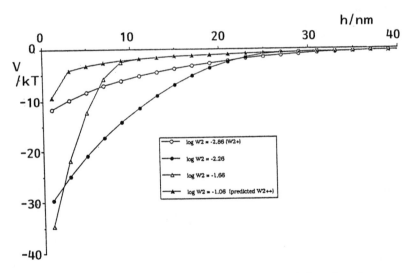

Fig. 11: V(h) plots for PS-g-PEO particles in the presence of PAA 14, at pH 9 and 10^{-1} mol dm^{-3} NaCl, for the various PAA wt. fractions indicated. Random-coil approximation.

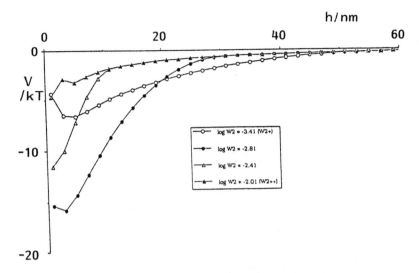

Fig. 12: V(h) plots for PS-g-PEO particles in the presence of PAA 14, at pH 9 and 10^{-2} mol dm^{-3} NaCl, for the various PAA wt. fractions indicated. Random-coil approximation.

As can be seen, at both 10^{-1} and 10^{-2} mol dm^{-3} NaCl, $V(h \to o)$ is smallest at $W_2^{\dagger}$ and $W_2^{\ddagger}$, and greater at intermediate PAA concentrations. Thus, at least semi-quantitatively, the pragmatic theoretical approach, outlined above, seems to account reasonably well for the experimental observations at these higher NaCl concentra-tions. When, however, these equations were applied at 10^{-3} or 10^{-4} mol dm^{-3} NaCl solutions, the corresponding V(h) curves showed no net attraction, clearly in contrast to the experimental observations (figs. 5 and 6)! The most probable explanation is that the use of eqn. (6) for $\Delta(\phi_2)$ is no longer valid, i.e. the random-coil approximation breaks down, at these low electrolyte concentrations. Figure 13 shows V(h) plots at these two lower electrolyte

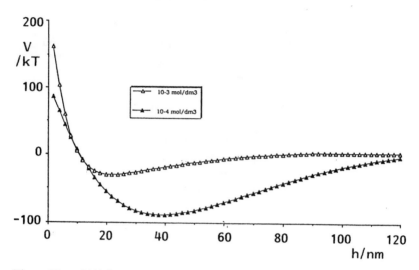

Fig. 13: V(h) plots for PS-g-PEO particles in the
presence of PAA 14, at pH 9, 10^{-3} or 10^{-4} mol dm^{-3} NaCl
and at $W_2^\dagger$. Rigid-rod approximation.

concentrations, using the same set of equations, except
that <u>no</u> dependence of Δ on ϕ_2 was now assumed (i.e.
$\Delta = \Delta_o = 1.4\ r_g$); as indicated earlier, this amounts to
assuming that the polyelectrolyte chains now behave as
rigid rods. The two V(h) curves in fig. 13 both show a
minimum in V(h), reflecting the balance of the repulsive
V_E and attractive (but longer-range) V_{dep} interations.
However, it seems as if the value of V_{min} is too large,
particularly at 10^{-4} mol dm^{-3} NaCl ($V_{min} \sim 80$ kT). It
would appear that the rigid-rod model overestimates V_{dep},
whilst the random-coil model underestimates V_{dep}, at <u>low</u>
background electrolyte concentrations. Therefore, although
the pragmatic model, based on eqn. (6) for $\Delta(\phi_2)$ works
reasonably well at <u>higher</u> electrolyte concentrations, clearly
a new approach for $\Delta(\phi_2)$ for polyelectrolyte at <u>low</u>

electrolyte concentrations is required.

It should be mentioned that Böhmer et al. (1990) have extended the Scheutjens-Fleer self-consistent, mean-field lattice model to the polyelectrolyte depletion situation. Some computated data from this approach are shown in fig. 14, for polyelectrolyte chains in an athermal solvent, at an uncharged, non-adsorbing ($\chi_s = 0$) surface. Only relatively high background electrolyte concentrations ($> 10^{-2}$ mol dm^{-3}) are considered.

Fig. 14(a) should be compared to fig. 10, wehre Δ has been calculated using eqn. (6) from the pragmatic theory. The general form of the curves in the two figures is similar. However, because of the mean-field assumptions in the Scheutjens-Fleer model, one difference is that Δ_o (limiting value of Δ at low ϕ_p) in fig. 14 (a) appears to be independent of electrolyte concentration. This seems unreasonable.

The form of the $\Delta A(M)$ plots given in fig. 14(b) may be compared with those for $V(h)$ given in figs. 11 and 12, calculated using the pragmatic theory for 10^{-1} and 10^{-2} mol dm^{-3} NaCl solution, respectively. One striking difference between the curves of Böhmer et al. (fig. 14(b), and those shown in figs. 11 and 12, is the presence of a maximum in the interaction energy curves in the former case at 10^{-1} and 10^{-2} mol dm^{-3} electrolyte. The origin of the repulsive contribution is related to the work required to displace the polyelectrolyte chains into bulk solution, against the Donnan (electrostatic) potential difference set up across the depletion zone between the particles and

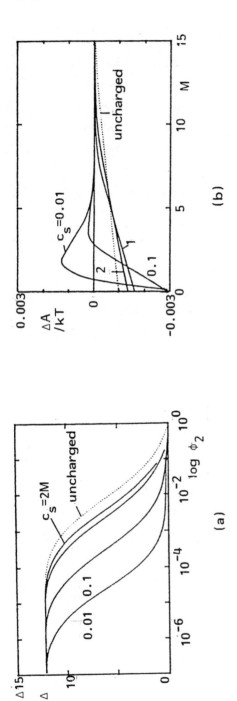

Fig. 14 (a) Depletion layer thickness Δ (lattice units) for a fully-charged polyelectrolyte, versus bulk volume fraction (ϕ_2);

(b) Interaction free energy ΔA (kT/segment area) versus plate separation M (lattice units), for $\phi_2 = 10^{-2}$.

In both cases the polyelectrolyte chain length is 500 segments, and $\chi = \chi_s = 0$; also the plate(s) are uncharged. The full curves represent four different 1:1 electrolyte concentrations as indicated; the dotted curve is for an uncharged polymer, for comparison.

the bulk polyelectrolyte solution. This potential difference is greater the lower the background electrolyte concentration. The presence of an energy barrier in the interaction curve implies some <u>kinetic</u> control of the depletion flocculation behaviour seen at low background electrolyte concentrations. However, we have no direct evidence for this; even in 10^{-4} mol dm^{-3} NaCl the flocculation seems to be thermodynamically controlled. Clearly, this is an aspect which needs further investigation; force-balance measure- would be useful here, if they can be made sufficiently sensitive.

CONCLUSIONS

It has been demonstrated that both bridging and depletion flocculation can occur rather when poly(acrylic acid) is added to aqueous dispersions of polystyrene latex particles, carrying terminally-anchored poly(ethylene oxide) chains, depending on the pH of the medium. In the case of depletion, the region of polyelectrolyte concentrations over which flocculation is observed, depends strongly on the background electrolyte concentration.

Two theoretical approaches to depletion flocculation have been examined. The first, a "pragmatic" approach, gives reasonable semi-quantitative agreement with the experimental data, at least at higher background electrolyte concentrations ($> \sim 10^{-2}$ mol dm^{-2}). It fails at lower electrolyte concentrations, at present, because no adequate theory as yet exists for predicting the depletion layer thickness, as a function of bulk polyelectrolyte concentration, at low salt concentrations. The second approach, based on the Scheutjens-Fleer theory of polymers at interfaces, also seems to lead to some unresolved questions, in particular, whether a significant energy barrier to flocculation exists, at low salt concentrations.

ACKNOWLEDGEMENTS

The authors would like to thank the SERC and I.C.I. Chemicals and Polymers Division plc for financial support. We also acknowledge many useful discussions with Dr. John Padgett of I.C.I.

REFERENCES

Auvray , L. (1981) J.Physique $\underline{42}$, 79.

Bailey, F.E.J. (1964). J.Polymer Sci. $\underline{192}$, 845.

Böhmer, M.R., Evers, O.A. and Scheutjens, J.M.H.M. (1990). Macromolecules $\underline{93}$, 2288

Boucher, E.A. and Hines, P.D. (1976). J.Polymer Sci. (Physics) $\underline{14}$, 2241.

Bromley, C. (1989). Colloids and Surfaces $\underline{17}$, 1.

Cawdery, N.F. (1992). Ph.D. thesis, Bristol University.

Cohen Stuart, M.A., Cosgrove, T. and Vincent, B. (1986). Advances in Colloid and Interface Science $\underline{24}$, 143.

Cosgrove, T. and Ryan, K. (1990). Langmuir $\underline{6}$, 136.

Cowell, C., Li-In-On, R. and Vincent, B. (1978). J.Chem. Soc. Faraday Trans. I $\underline{74}$, 337.

Cowell, C. and Vincent, B. (1982). J.Colloid Interface Sci. $\underline{87}$, 518.

Doroszkowski, A. (1986). Colloids and Surfaces $\underline{17}$, 13.

Emmett, S. and Vincent, B. (1990). Phase Transitions $\underline{21}$ 197.

Evans, R. and Napper, D.H. (1973). Nature $\underline{246}$, 34.

Fayt, R., Forte, R., Jacobs, C., Jerome, R., Ouhadi, T., Teyssie, Ph. and Varshney, S.K. (1987). Macromolecules 20, 1443.

Fleer, G.J., Scheutjens, J.H.M.H and Vincent, B. (1984). In Polymer Adsorption and Dispersion Stability (Ed. E.D. Goddard and B. Vincent) American Chemical Society 240, 245.

Ikawa, T., Abe, K., Honda, K., Tsuchida, E. (1975) J.Polymer Sci. A13, 1505.

Kawaguchi, Y. and Nagasawa, A. (1969). J.Phys.Chem. 73, 4382.

Klein, J.W., Gnanou, Y. and Rempp, A. (1990). Polymer Bull. 24, 39.

MacMillan, R. (1989). Ph.D. thesis, Bristol University.

Miller, M.L. and Rauhut, C.A. (1958). J.Amer.Chem.Soc. 80, 4115.

Miller, M.L., Botty, M.C. and Rauhut C.A. (1960). J.Colloid Sci. 15, 83.

Milling, A., Vincent, B., Emmett, S., Jones, D.A.R. (1991). Colloids and Surfaces 57, 185.

Napper, D.H. (1983). Polymeric Stabilization of Colloidal Dispersions, Academic Press, London.

O'Brien, R.W. and White, L.R. (1978). J.Chem.Soc. Faraday Trans. II 74 1607.

Rawson, S., Ryan, K. and Vincent, B. (1988). Colloids and Surfaces 34, 89.

Snowden, M.J., Clegg, S.M., Williams, P.A. and Robb, I.D. (1991). J.Chem.Soc. Faraday Trans. 87, 2201.

Takahashi, A. and Nagaswa, M. (1964). J.Amer.Chem.Soc. 86 543.

Vincent, B., Edwards, J., Emmett, S. and Jones, D.A.R. (1986). Colloids and Surfaces, 17, 261.

Vincent, B. (1990). Colloids and Surfaces 50, 241.

SOME GENERAL OBSERVATIONS ON THE INTERACTION OF NATURAL RUBBER LATEX PARTICLES WITH TIN TAILING SLIMES

C.C. HO[1], K.C. LEE[1], E.B. YEAP[2]

[1]Department of Chemistry and [2]Department of Geology, University of Malaya, 59100 Kuala Lumpur, Malaysia

ABSTRACT

Tin tailing slimes from ex-mining ponds consist mainly of sand, silt and clay minerals. The larger size fractions of sand and silt sediment relatively fast whereas the very fine clay particles form a very stable colloidal dispersion which is responsible for the muddy look of such a slurry in more dilute form. The interaction of these clay particles with natural rubber (NR) latex particles in a mixed colloid system, in which both types of particles are negatively charged and comparable in size, is studied through the adsorption of latex particles on the clay minerals. It was found that the adsorption of NR latex particles on the clay particles was almost spontaneous and the extent of adsorption depends strongly, not only on the clay mineral composition and the pore fluid chemistry of the slimes, but also the composition of the latex. This strong affinity of the latex particles to the clay was attributed to the 'electrostatic bridge' made possible by the presence of significant amount of soluble and exchangeable divalent cations in the slimes. The presence of high non-rubber contents can, on the other hand, give rise to competitive adsorption with the latex particles for the clay surface. For comparison, pure kaolinite and montmorillonite in the sodium form in various proportions to simulate the clay composition of the slimes was employed. As expected very little adsorption took place on this clay surface. Transmission electron microscopy (TEM) results of the NR latex-slime mixtures revealed that latex particles were adsorbed on the edges of the clay particles. This correlates well the high affinity and very small amount of latex adsorbed as observed above.

INTRODUCTION

The colloid stability of disperse systems containing only one type of particles is well-studied and fairly well-understood via the DLVO theory [1,2]. The interaction of mixed colloids, with more than one type of particle, is expected to be more complex, and more sophisticated models are required. For example, Hogg et al [3] first worked out quantitatively the interaction of two dissimilar electrical double layers. This has been extended theoretically by Usui [4]. Experimentally, the mixed colloids used could make up of particles of completely different materials but of comparable sizes [5], or of system containing similar particles where the particle size differed greatly [6]. Yet another variation of the mixed-colloid systems are those of unlike materials with different particle sizes [7]. The interaction of particles of greatly differing sizes is usually monitored via the adsorption isotherms of the small particles

on the larger ones [8]. Thus in principle the
problem is closely related to the adhesion of
small particles on extended surfaces.

The tailing left behind after the mining of tin
ore is a remarkably stable mineral suspension
called slime or slurry. This is normally held
back in a mined out area, forming what is known
as the ex-mining pond. The larger fractions of
sand and silt particles in the slime would
sediment relatively fast. On the other hand, the
very fine clay mineral particles would remain
dispersed in suspension, giving the muddy look
of such a slurry in its dilute form. The clay
minerals consist mainly of kaolinite, illite,
some small amount of montmorillonite and
amorphous materials. These particles were all
negatively charged under conditions in which the
slimes were found (pH ca 4.0-7.0). On the other
hand, natural rubber (NR) latex concentrate
obtained from *Hevea brasiliensis* is essentially
a dispersion of polyisoprene particles in an
aqueous serum phase containing a small amount of
soluble, non-rubber constituents. These latex
particles were stabilised by adsorbed proteins
and adsorbed long chain fatty acid soaps derived
from the hydrolysis of phospholipids present
originally in the latex from the tree [9]. Thus
the latex particles are amphoteric in nature and
become negatively charged at alkaline pH [10].

While the interaction between synthetic polymer
latex particles and metal oxide particles has
been reported in a few studies [7,8], the study
of the interaction between clay and latex particles as a mixed colloid system does not seem to

be available in the literature. The closest was
the flocculation of kaolinite and quartz particles by an 'associated colloidal flocculant', an
almost water-insoluble polymer dispersed in the
form of relatively large colloidal particles
through molecular association [11]. Thus it
would be of interest to investigate the adsorption of natural rubber latex particles on the
clay minerals of the slime samples and to see
how different this type of interaction differs
from analogous interactions between clay particles and, say, polymer [12], surfactant [13] and
bioflocculant [14]. This should have direct
bearing on the practical applications of such
'particulate assemblage', for example in the
dewatering of clay slurry. This paper presents
some preliminary findings on three types of
slimes with different clay compositions. Some
salient features of the adsorption of NR latex
particles on these at pH 7, where both types of
particles are negatively charged, are discussed.
The adsorption at this pH was also chosen
because this was the natural pH in which most of
the local slimes were found.

EXPERIMENTAL

Materials

The slimes were collected from three different
ex-tin mining slurry ponds located respectively
at Malim Nawar (MN), Pengkalan (PN) and Serendah
(SH), Malaysia. The slimes were air-dried,
pulverized and sieved (60 mesh BS) to obtain
fine clay samples which were stored at ambient

temperature before use. For electrophoresis work
these sieved slime samples were further cleaned
by washing with distilled water to remove
inorganic and other water-soluble impurities.
1.0 g of sample was dispersed in 50 ml of
distilled water, shaken for 20 min, centrifuged
to remove the supernatant and the process
repeated by redispersing the sediment in 50 ml
of distilled water. The washing was repeated
until the conductivity of the supernatant
approached that of the distilled water used.

The sodium forms of the kaolinite and montmorillonite (BDH Chemicals) clays were prepared by
the recommended method [15]. The dry clays were
dispersed in 1.0 mol dm^{-3} NaCl solution to give
a 4.0% (w/v) dispersion which was equilibrated
with agitation for 24 h. It was then centrifuged
and the supernatant removed. The sediment was
then redispersed in another fresh portion of
NaCl and the process of equilibration and
centrifugation repeated three more times. The
excess chloride ions were removed by repeated
washing of the clay with distilled water. The
cleaned clay was air-dried and ready for use.

Commercial high-ammonia NR latex concentrate
(HA) prepared by centrifugation and an evaporated NR latex concentrate (EL) from Revertex
Ltd were used for this study. The dry rubber
content (DRC) of HA latex was about 60% (w/w)
and that for EL was 71% (w/w). They were used
without further purification. Polystyrene latex
was prepared by surfactant-free emulsion polymerization [16] followed by extensive dialysis of
the latex before use. Electron microscopy gave

TABLE 1

Some physical and chemical parameters of NR latex concentrate

	HA	EL
Ash content % (w/w)	0.52	8.51
D.R.C. % (w/w)	60.5	70.0
T.S.C. % (w/w)	61.0	71.0
Density (g/cm^3)	0.913	0.913
Particle size (μm)	0.05 - 0.60	-
	av. 0.24 $\pm$ 0.13	-
Mineral content (mg/g dry rubber)		
Mg	140	799
Zn	390	555
Cu	3	2
Fe	6	9
Ca	15	22
K	787	937

an average particle diameter of 0.420 μm for this latex. Some parameters of the NR latices are given in Table 1.

Adsorption studies

4.0 g of the treated slime sample were dispersed in 50 ml of distilled water, followed by 10.0 ml of electrolyte solution at an appropriate concentration. The clay dispersion was then adjusted to the desired pH with hydrochloric acid or ammonium hydroxide. Weighed amount of HA

latex concentrate was first diluted with 35 ml
of distilled water followed by pH adjustment.
Initially the clay dispersion, contained in a
100 ml measuring cylinder, was inverted by hand
15 times. The latex dispersion was then added
followed by distilled water to give a final
volume of 100 ml. Immediately the cylinder was
again inverted another 15 times. The total time
taken for the inversion procedure was ca 2 min.
The dispersion in the measuring cylinder was
allowed to equilibrate at $29^{o}C$ (room temperature) for 40 min. Then 20 ml of the dispersion
was withdrawn, using a pipette, from below the
liquid surface in the measuring cylinder. This
was lightly centrifuged at 2500 rpm for 20 min
to remove large clay particles without causing
the creaming of the free latex particles. The
supernatant was carefully withdrawn from each
centrifuge tube and its turbidity measured. The
concentration of latex particles was deduced
from a calibration curve obtained by measuring
the turbidity of latex dispersions containing
different solid contents after lightly centrifuging as described above. The amount of latex
adsorbed was determined, after a constant equilibration time, to obtain the adsorption isotherm and at different equilibrating times to
evaluate the adsorption rate. The rate experiments showed that the uptake of latex was almost
immediate and thereafter remained constant up to
3 h. Consequently an equilibrating time of 40
min was fixed for establishing the adsorption
isotherms. Longer equilibrating periods were
avoided to prevent homocoagulation of latex
particles especially at pH<7. It was also noted
that adsorption/adhesion of latex particles on

the walls of the measuring cylinder was negligible.

Measurement of electrophoretic mobility

2.0 mg of the treated slime were dispersed in 50 ml of electrolyte solution at the appropriate concentration and shaken for 3 h. Equilibrium studies showed that there was no further change in electrophoretic mobility after 2 h [17]. The mobility of the slime dispersion was determined using a Rank Brothers Particle Microelectrophoresis apparatus at $25.0^{\circ}C$. An average of at least 20 readings of particle mobility was determined for each dispersion. The direction of current flow was reversed after each reading to eliminate polarisation of the electrodes. In the case of latex dispersion, a stock dispersion containing $4\times10^{-3}\%$ (w/v) solids was prepared from the HA latex concentrate by dilution with distilled water. To prepare a dispersion for mobility measurement, 0.15 ml of the stock dispersion was dispersed in 50 ml of 1.5×10^{-2} mol dm^{-3} sodium chloride solution and the pH adjusted. The mobility was measured 10 min after mixing and studied as a function of pH.

Electron microscopy

All NR latices were diluted and fixed with a 2% solution of osmium tetroxide prior to deposition onto collodian films supported on copper grids before observation in the TEM under low beam current. Samples of NR latex-slime aggregates were taken from the sediment in the measuring cylinder from the adsorption experiment mentioned above, with minimum disturbance, and were diluted slightly with a few drops of the clear supernatant from the same sample and then similarly fixed with osmium tetroxide before observation. No staining with osmium tetroxide was required for the samples of slime with adsorbed polystyrene latex. These were deposited onto the grids and observed directly. Samples of slime dispersions at pH 7 were observed directly after deposition on grids to minimize the effect of aggregation. Very dilute dispersions were used.

RESULTS AND DISCUSSION

Nature of slimes

The three slimes used in this work have been characterised by X-ray diffraction, chemical analysis, particle size analysis and electron microscopy [18]. These techniques established that PN slime had a higher percentage of highly-crystalline kaolinite than MN slime, in addition to having about 8% of montmorillonite, while SH slime had the highest amorphous material content (24%). Details of the clay mineralogy of the slimes and some chemical and physical parameters are given in Table 2. Table 3 shows the exchangeable cations and composition of pore fluid of the slimes.

Electrophoretic mobility

The electrophoretic mobility versus pH curves for the HA and EL latices in the presence of 1.5×10^{-2} mol dm^{-3} NaCl are shown in Fig 1. The

TABLE 2

Mineral composition and physical parameters of slimes

| Clay (%, w/w) | Samples | | |
	MN	PN	SH
Kaolinite	65.1	71.2	32.3
Illite	10.2	8.2	9.7
Amorphous clay	10.0	nd	24.7
Montmorillonite	nd	7.8	nd
Quartz	15.3	12.0	33.3
Amorphous content*	5.6	1.4	20.0
Density (g/cm^3)	2.6	2.6	2.6
Particle size (μm)			
range	0.10-0.88	0.12-0.35	–
average	0.38±0.18	0.17±0.06	

* – determined by chemical analysis

respective isoelectric point (IEP) was 4.25 for HA latex and 3.45 for EL latex. Below this pH the latex particles were positively charged. Above the IEP the negative mobilities of the latex particles increased rapidly with increasing pH until it reached a constant value of -3.53×10^{-8} m^2 s^{-1}v^{-1} at pH 7.25 for HA latex and -3.36×10^{-8} m^2 s^{-1}v^{-1} at pH 7.00 for EL latex. pK$_a$ values of 5.55 and 4.95 were deduced [19] for the surface ionogenic groups on HA and EL latex particles respectively. These values were slightly higher than those obtained previously [9], in which the latex had been extensively dialysed and most of the smaller particles eliminated. The maximum and constant mobilities

TABLE 3

Exchangeable cations and pore fluid compositions of slimes

Exchangeable cations (meq/100g dry slime)	Samples MN	PN	SH
Ca^{2+}	3.71	26.4	1.21
Mg^{2+}	0.53	0.97	0.01
K^+	0.10	0.06	0.06
Na^+	nd	nd	0.11
Al^{3+}	0.97	0.90	nd
Pore fluid composition: Soluble cations (meq/100g dry slime)			
Ca^{2+}	0.35	0.45	0.10
Mg^{2+}	0.08	0.06	<0.01
K^+	0.05	0.10	0.02
Na^+	nd	nd	0.04
Pore fluid composition: Soluble anions (meq/100g dry slime)			
Sulphate	46.2	126.0	nd
Chloride	88.7	110.0	nd
Bicarbonate	12.8	45.1	1.0
Carbonate	nd	nd	nd

nd - not detected

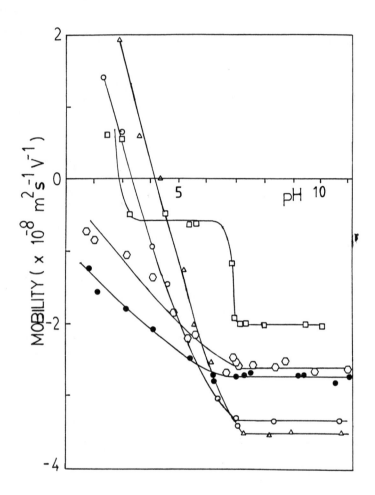

Fig.1   Variation of electrophoretic mobility
with pH for: EL (o) and HA (Δ) latices in
1.5×10^{-2} mol dm^{-3} NaCl; and SH (□), PN (●) &
MN (◯) in 5×10^{-2} mol dm^{-3} NaCl.

of the present systems were also much lower than
those obtained previously [9]. However, the
shape of the mobility vs pH curves is similar.
This shows that the soluble constituents, both
organic and inorganic, originally present in the
latices from the trees exert a strong influence
on the surface charge of the latex particles.
The surface charge of the latex particles is
derived from adsorbed long chain fatty acid
soaps and adsorbed proteins [10]. Adsorption of
the latex particles on clay was thus carried out
at pH 7 where both types of particles are fully
ionised and negatively charged.

The electrophoretic mobility behaviour of the
clay particles in the slimes in the presence of
5×10^{-2} mol dm^{-3} NaCl has been described
previously [17] and the mobility – pH curves for
the three slimes are also shown in Fig 1 for
easy reference.

Separation efficiency

Since the adsorption of NR latex particles on
clay particles in the slime was deduced by
turbidity measurement of the supernatant after
separation of clay particles, the efficiency of
the separation by centrifugation was checked to
ensure complete separation was achieved. The
density of the NR latex particles is 0.913 g/cm^3
while that for the clay minerals is 2.6 g/cm^3.
Thus under normal conditions of stable
dispersion, NR latex particles would cream very
slowly and clay sediments but rather slowly
under gravity. A 4% (w/v) clay dispersion
containing 0.2% (w/v) HA latex would, if left

undisturbed, appear brownish in colour and
remain stable without sedimentation of the clay
particle for at least one week. After being
centrifuged lightly at 2500 rpm for 20 min, the
dispersion would become whitish in colour (see
Fig 2). This clearly shows that all the clay
particles can be brought down at this speed,
leaving behind all the unaggregated HA latex
particles in dispersion. A higher g force was
not used in order to avoid homocoagulation or
creaming of the latex particles themselves. Such
a light centrifugation technique had been used
previously [8] in the separation of polystyrene
latex particles from ferric hydroxides and
aluminium hydroxides. The efficiency of the
separation would decrease with increase in
concentration of unaggregated latex particles in
the dispersion after adsorption.

Effect of equilibration time

Fig 3   shows that the adsorption of   HA latex
particles on the clay particles at pH 7 in the
presence of 1.7×10^{-3} mol dm^{-3} NaCl was almost
spontaneous even though both clay and latex
particles are negatively charged at this pH. The
maximum amount of latex adsorbed after 40 min of
equilibration time was 16% (w/w) from a 0.1 %
latex dispersion on MN slime whereas 22% (w/w)
of latex were adsorbed from a 0.2% latex
dispersion on PN slime. This works out to be
about 3.1×10^{12} latex particles adsorbed per
gram of slime for MN slime and 8.6×10^{12} for PN
slime. In terms of the number of latex particles
adsorbed to total clay particle number present
in the system, ratios of about 23:100 for MN

Fig.2 Separation efficiency by centrifugation
for slime-latex mixture containing 4% (w/v) PN
slime and 0.2% (w/v) HA latex in the presence of
8.56×10^{-4} mol dm^{-3} NaCl at pH 7.
(1) before centrifugation;
(2) after centrifugation.

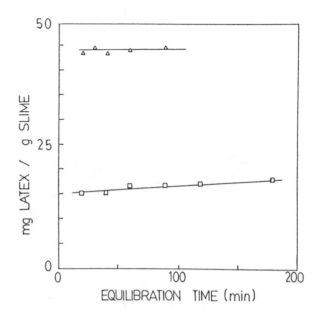

Fig. 3. The effect of equilibration time on adsorption of HA latex on PN (△) and MN (☐) slimes in the presence of 1.7x10⁻³ mol dm⁻³ NaCl at pH 7. Weight of slime was 4.0% (w/v). 0.2% (w/v) HA latex was used for PN and 0.1% (w/v) for MN slimes.

slime and 9:100 for PN slime respectively were obtained. The ratio of the latex particles originally present before adsorption to total clay particles was about 37:100 for MN slime and 11:100 for PN slime. Thus, based on the initial latex particle number present, the number of latex particles adsorbed was in excess of 60%, but in terms of absolute weight of latex adsorbed this was extremely small compared to the amount of slime used in the system. 4% (w/v) slime dispersions were used in both cases.

There was only a very marginal increase in the amount of latex adsorbed when the equilibration time was increased from 20 min to 3 hours for the MN slime, whereas there was no change in the amount adsorbed even up to 100 min for PN slime. Thus, even though the actual amount of latex adsorbed was extremely small, the affinity of the latex for the clay was surprisingly high. The rate of adsorption of negatively charged polystyrene latex on positively charged ferric and aluminium hydroxides [8] was very much slower in comparison. In the present system, the amount of latex adsorbed on the clay particles would be expected to be very small, since both types of particles are negatively charged and hence would experience strong repulsion between them. In addition the monovalent electrolyte used, Na^+ ions, would not be effective enough to compress the double layers to reduce this repulsive force, neither is it capable of functioning as an 'electrostatic bridge' for linking the latex to the clay surface. Even though the net charge of the clay particles was negative, there would still be some residual cationic sites at the edges of the clay particles. Thus it is thought that adsorption of latex particles is through the small number of residual positive charge at the edges of the clay particles. This is borne out by TEM results shown later.

Effect of slime concentration

The adsorption isotherms of HA latex on MN slime at four different slime concentrations and at pH 7 are shown in Fig 4. Straight line plots were obtained for these and the slopes of the lines

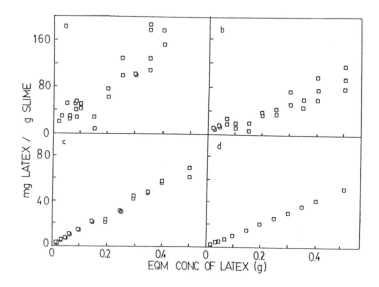

Fig.4 Effect of MN slime concentration on the Henry's constant in the adsorption of HA latex in the presence of 8.56×10^{-4} mol dm^{-3} NaCl at pH 7 : (a) 0.5% (w/v), (b) 1.0% (w/v), (c) 4.0% (w/v) and (d) 8.0% (w/v) slimes.

TABLE 4

Effects of MN slime concentration on Henry's constant in the adsorption of HA latex in the presence of 8.56×10^{-4} mol dm^{-3} NaCl at pH 7

Weight of slime (g)	K (per g)	Correlation coefficient, r
0.5	0.456 ± 0.057	0.80
1.0	0.196 ± 0.017	0.86
4.0	0.140 ± 0.003	0.99
8.0	0.109 ± 0.001	1.00

(designated as Henry's constants, K) decreased
with increasing slime concentration. A four-fold
decrease in K values was noted when the slime
concentration was increased from 0.5 g to 8.0 g
(Table 4). As the concentration of slime is
increased, the clay surface area available for
adsorption would increase. Thus, the amount of
latex adsorbed per gram of slime would decrease.
At the same time the particle number ratio of
latex to clay would also decrease. Both these
effects would result in a decrease in the
Henry's constant as the slime concentration was
increased. There was considerable scatter in
the adsorption data at low concentration of
slime but the accuracy of the adsorption
measurement improved with increasing slime
concentration. On the other hand too large a
latex to clay ratio would reduce the efficiency
of separation of latex from the slime because of
interference of the large number of unadsorbed
latex particles. Thus an optimal slime
concentration would appear to be 4 % (w/v).

Effect of slime and clay types

In addition to the slimes, pure kaolinite and
montmorillonite mixed in different proportions
were used to roughly simulate the clay
composition of the slime in the adsorption of HA
latex. The adsorption isotherms obtained using 4
g of slime or pure clay sample at pH 7 in the
presence of 8.56×10^{-4} mol dm^{-3} NaCl were again
straight lines where the values of Henry's
constant for PN slime was the largest whereas
that for MN slime was the smallest with that for
SH slime intermediate (Fig 5). The Henry's

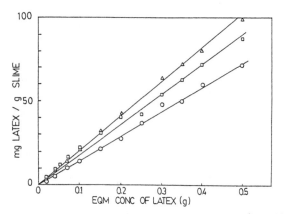

*Fig.5   Effect of slime types on the Henry's
constant in the   adsorption of HA latex in the
presence of 8.56x10^{-4} mol dm^{-3} NaCl at pH 7.
Weight of slime 4.0% (w/v). (O) MN, K=0.144/g;
(△) PN, K=0.212/g; (□) SH, K=0.191/g.*

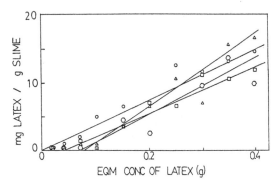

*Fig.6 Effect of clay types on the Henry's
constant in the   adsorption of HA latex in the
presence of 8.56x10^{-4} mol dm^{-3} NaCl at pH 7.
Weight of clay mixture 4.0% (w/v). △ Kaolinite
alone, K=0.051/g; ◐ [92% (w/w) kaolinite + 8%
(w/w) montmorillonite], K=0.045/g; O [80% (w/w)
kaolinite + 20%(w/w) montmorillonite], K=0.036/g;
  □ [70% (w/w) kaolinite + 30% (w/w) montmoril-
lonite], K=0.030/g.*

constants were very much smaller for pure kaolinite and the various mixtures with montmorillonite as shown in Fig 6. In the case of pure kaolinite, no adsorption of latex took place below 0.07 g latex.

PN slime contained the highest level of exchangeable cations and soluble cations (see Table 2 & 3) among the three slimes used. These cations especially Ca^{2+} and Mg^{2+} could become 'electrostatic bridge' binding latex particles to the clay. Such phenomena has been observed before where the presence of Ca^{2+} resulted in enhanced adsorption of anionic bioflocculant on bentonite [14] and of anionic surfactant on kaolinite [13]. On the other hand, aluminium hydrolyzable cations could function as effective 'anchor points' for anionic polymer in the flocculation of clay minerals [20]. In addition divalent cations could effectively reduce the zeta potential of the clay minerals [18] and hence the repulsion between latex and clay particles resulting in an increase in adsorption of the latex. It has been found previously [21] that the free energy of adsorption of small, positive polystyrene latex particles onto much larger, negative polystyrene particles, both with adsorbed poly (vinyl alcohol) (PVA) molecules, was reduced in the presence of electrolyte. In view of these points, the adsorption of latex on PN slime would be enhanced leading to a larger K value.

A slightly smaller K value was obtained for SH slime which contained high amorphous Fe_2O_3 content. Most of these amorphous materials exist

in the form of metal hydroxide at pH 7 [17] and
thus could not function as 'electrostatic
bridge'. However the electrophoretic mobility
(and hence the surface charge) of the SH slime
was also the lowest amongst the three slimes
[17]. Thus a balance of these two opposing
effects would result in an affinity for latex
particles intermediate between those of PN and
MN slimes.

In the case of the pure kaolinite and montmorillonite, the adsorption isotherms were all
straight lines but the amount of latex adsorbed
on them was very small compared to the natural
slimes and hence significantly lower K values
were obtained for these systems. This clearly
shows that the soluble and exchangeable cations
in the natural slimes, especially the divalent
ones, have a profound influence on the latex
adsorption behaviour. Without these cations, the
pure clay systems in the sodium form would be
unable to form sufficient 'electrostatic
bridges' for adsorption to occur and hence the
affinity of the latex particles for these clay
surfaces would be very small. Furthermore cationic sites on the broken edges of the clay
minerals are known to represent only 2 - 5 % of
the total clay surface [22], thus the number of
sites available for latex adsorption is extremely small. In comparison, weathering conditions
are known [23] to form more disrupted clay
minerals which could increase the number of cationic sites on the clay particles in the slimes.
Therefore the amount of latex adsorbed on the
pure clay surface became measurable only after
certain equilibrium concentration of latex has

been reached. This then explains the induction region observed with the pure clay system below which concentration no adsorption takes place.

Effect of latex type

EL and HA latices were used for a comparison study. The EL latex, being manufactured by evaporation, was characterised by a higher non-rubber solids and soluble ions than those for HA latex [24]. The shape of the adsorption isotherms of the these latices on PN and MN slimes was again straight line (Fig 7). The K values of HA (0.143) and EL (0.140) latices on MN slimes were about equal. However, that for EL (0.17) latex was ca 19% smaller than that on HA latex (0.21) in the case of adsorption on PN slime. This means a lower affinity of EL latex particles for the PN slime. Since the electro-phoretic mobility of the EL latex particles was slightly lower than that of the HA latex over the pH range 7 to 10, a larger amount of EL latex was expected to be adsorbed on PN slime compared to HA latex. That this was not observed was attributed to the higher non-rubber contents of the EL latex. These non-rubbers (such as pro-teinaceous materials, higher fatty acid soaps, salts of non-volatile acids and carbonates/bi-carbonates) of El latex could become adsorbed on the clay particles upon mixing the two. The higher amount of soluble and exchangeable cations, in particular Ca^{2+} and Mg^{2+}, could also coordinate with these biological materials. Thus the cationic sites or 'electrostatic bridges' on the clay available for latex adsorption would be reduced as a result of the competition between

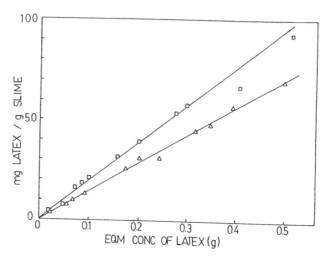

Fig.7 Adsorption isotherm of EL latex on 4.0%
(w/v) MN (△) [K=0.140/g] and PN (□) slimes
[K=0.173/g] in the presence of 8.56x10^{-4} mol
dm^{-3} NaCl at pH 7.

the latex particles and biological materials
for the clay surface. This did not occur in the
case of MN slime.

A complete adsorption isotherm for polystyrene
latex on the slime was not carried out but
instead the amount of latex adsorbed from a 0.1%
(w/v) latex dispersion at pH 7 was checked using
8.0 g of MN slime in the presence of 8.1 x 10^{-4}
mol dm^{-3} (90 ppm) of calcium chloride. Only a
very small amount of the latex was adsorbed
(1.81 mg of polystyrene latex/g of slime). Under
similar condition, 12.4 mg of HA latex/g of
slime was found to be adsorbed on MN slime.
However when PN slime was used but in the
presence of 1.8 x 10^{-3} mol dm^{-3} (200 ppm) of
calcium chloride, no adsorption of polystyrene

latex on the slime was observed at all. The
corresponding situation showed an adsorption of
12.4 mg of HA latex on PN slime. This certainly
indicates a much lower affinity of the polystyrene latex particles for the slime compared with
the NR latex particles. This behaviour was bro-
adly confirmed by the TEM results given below.

Electron microscopy study

The occurrence of disrupted kaolinite platelets
and halloysite minerals are rather common in
slime samples which are well-weathered as
depicted in Fig 8a for MN slime. This contrast
strongly with the pure clay systems. Furthermore
the clay particles of the slimes were not
completely dispersed to begin with, even though
very dilute dispersion was used. Some aggregates
were present even at pH 7. The TEM specimens of
the NR latex-slime aggregates used appeared
rather thick compared to ordinary TEM samples as
constrained by the preparation method. Micrograph in Fig 8b clearly showed the adsorption of
small latex particles on the edges of kaolinite
platelets. Since the particle size distributions
of the NR latex and the slime were both broad,
the adsoprtion of some small clay particles on
the larger latex particles was also observed
(Fig. 8a & c). This bears close resemblance to
the PVC latex particles coated with Ludox HS
silica particles as reported previously by
Matijevic and co-worker [25]. It should be noted
that in the present system, the number of latex
particles adsorbed on a clay particle is very
small indeed. On the other hand there seems to
be slightly more tiny clay particles adsorbed on

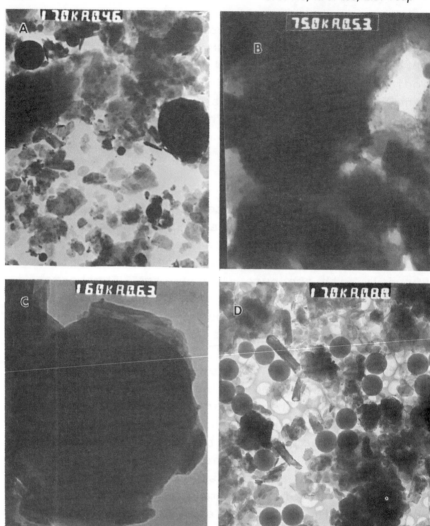

Fig.8 Transmission electron micrographs of slime-latex mixtures: (a) 8% (w/v) MN, 0.1% (w/v) HA latex in 8.11×10^{-4} mol dm^{-3} $CaCl_2$ pH 7; (b) 8% (w/v) PN, 0.1% (w/v) HA latex in 1.8×10^{-3} mol dm^{-3} $CaCl_2$ pH 7; (c) 8% (w/v) SH, 0.1% (w/v) HA latex at pH 7; (d) 8% (w/v) PN, 0.1% (w/v) polystyrene latex in 1.8×10^{-3} mol dm^{-3} $CaCl_2$ pH 7

larger size latex particles. The ratio of the number of latex particles to clay particles in the present system is very small compared to other mixed colloid systems where the particles are oppositely charged as reported in the literature [7,8,21 & 26]. However this small number of latex particles associated with the clay particles seem sufficient to cause aggregation of the slime resulting in heterocoagulation of the system. This is likely the reason why attempts to view the NR latex-slime aggregates under the scanning electron micrscope did not seem to reveal any latex particles at all except occasional ones. Polystyrene latex particles did not adsorb on the larger clay particle surface under similar condition (Fig 8d) but appeared to be distributed loosely and randomly among the clay particles. This is perhaps the reason for the inability of polystyrene latex to destabilize the slime dispersions.

CONCLUSIONS

Well-defined metal oxides and synthetic polymer latex particles and their appropriate combinations are often used as models of mixed colloid systems in the investigation of heterocoagulation pheonomena. Discrete particles were involved and the mechanism is one of adsorption of the smaller particles on the larger ones. On the other hand, water-soluble polymers have long been employed as flocculants for destabilizing dispersions such as in water clarification. An adsorption mechanism is again operational. The present study on a practical system between the

interaction of a slime dispersion and NR latex
particles is interesting in that it illustrates
the possible application of NR latex as a flocculant in the dewatering of clay dispersions and
points to a close similarity with a heterocoagulation process. Under the experimental conditions, the particles of the two dispersions
carried charges of the same sign and were of
comparable size. In contrast to well defined
model systems, the slimes and the NR latex are
complex, with respect to composition and polydispersity in particle sizes. The complexity of
this is clearly demonstrated by TEM in which it
is seen that small negative clay particles can
adsorb on large negative latex particles and so
can small negative latex particles adsorb on
large negative clay surfaces. And yet the
results could be interpreted reasonably well, at
least qualitatively at this stage, by known
colloidal behaviour such as heterocoagulation.
More detailed studies on these systems are in
progress.

ACKNOWLEDGEMENT

Financial support by the International
Development Research Centre (IDRC), Canada is
gratefully acknowledged.

REFERENCES

1. Derjaguin, B.V. and Landau, L.D. (1941). *Acta
   Physicochem URSS*, **14**, 633.

2. Verwey, E.J. and Overbeek, J.Th.G. (1948). *Theory of the Stability of Lyophobic Colloids*, Elsevier, Amsterdam, p193.
3. Hogg, R., Healy, T.W. and Fuerstenau, D.W. (1966). *Trans. Faraday Soc.*,, **62**, 1638.
4. Usui, S. (1973). *J. Colloid Interface Sci.*, **44**, 107.
5. Bleier, A. and Matijevic, E. (1976). *J. Colloid Interface Sci.*, **55**, 510.
6. Harding, R. D. (1972). *J. Colloid Interface Sci.*, **40** 164.
7. Dumont, F., Ameryckx, G. and Watillon A. (1990). *Colloids & Surfaces*, **51**, 171.
8. Hansen, F.K. and Matijevic, E. (1980). *J.C.S. Faraday I*, **76**, 1240.
9. Ho, C.C. and Ng, W.L. (1979). *Colloid & Poly. Sci.*, **257**, 406.
10. Ho, C.C. (1989). *Colloid & Poly. Sci.*, **267**, 643.
11. Kashiki, I and Suzuki, A. (1986). *Ind. Eng. Chem. Funda.*, (a) **25**, 120; (b) **25**, 444.
12. Dollimore, D. and Horridge T.A. (1973). *J. Colloid Interface Sci.*, **42**, 581.
13. Poirier, J.E. and Cases, J. M. (1991). *Colloids & Surfaces*, **55**, 333.
14. Levy, N., Bar-Or, Y. and Magdassi, S. (1990). *Colloids & Surfaces*, **48**, 337.
15. Laboratory Manual for Soil Analysis, Geotechnical Research Centre, McGill University, Montreal, Canada (1988).
16. Goodwin, J.W., Hearn, J., Ho, C.C. and Ottewill R.H. (1973). *British Polymer J.*, **5**, 347.
17. Ho, C.C., Lee, K.C. and Yeap, E.B. (1992). *Colloids & Surfaces*, (submitted for publication).

306 C.C. Ho, K.C. Lee, E.B. Yeap

18. Lee, K.C., to be published.
19. Ottewill, R.H. and Shaw J.N. (1967). *Kolloid Z. u. Z. Polym.*, 218, 34.
20. Roberts, K., Kowalewska, J. and Friberg, S. (1974). *J. Colloid Interface Sci.*, 48, 361.
21. Vincent, B. and Young, C.A. and Tadros, T.F. (1978). *Faraday Disc Chem. Soc.*, 65, 296.
22. Hanna, H.S. and Somasundaran, P. (1979). *J. Colloid Interface Sci.*, 70, 181.
23. Townsend, W.N. (1973)). *An Introduction to the Scientific Study of Soil*, 5th edi., Edward Arnold, London, p 15-23.
24. Calvert, K.O. (edi) (1982). *Polymer Latices and Their Applications*, Applied Science, London, p17.
25. Bleier, A. and Matijevic, E. (1978). *J. Chem. Soc. Faraday 1*, 74, 1346.
26. Furusawa, K. and Anzai, C. (1987). *Colloid & Polymer Sci.*, 265, 265.

NONIONIC SURFACTANT ADSORPTION IN LATEX, OXIDE AND CARBON BLACK COLLOIDAL SYSTEMS

Jeffrey R Aston[†], Peter J Scales, Janine S Godfrey[¶] and Thomas W Healy

Advanced Mineral Products Research Centre, School of Chemistry,
University of Melbourne, PARKVILLE, 3052, AUSTRALIA

ABSTRACT

The adsorption of alkyl phenol ethoxylate, polydistributed surfactants onto a range of solid surfaces produces very similar isotherms at high coverage in all cases. The adsorption reflects the same hydrophobic and hydrophilic forces that drive self-assembly of these surfactants in solution. This suggests that at high coverage, the solid-liquid interface is a site of nucleation of self-assembled interfacial objects.

At low coverages, however, the surfactant isotherms are either high affinity (carbon black, hydrophobic silica, low charge latex) or low affinity S-shaped co-operative isotherms (silica, high charge latex).

These isotherm characteristics have been examined in terms of the structural properties of the nonionic surfactants and the interfaces concerned. The essential differences between isotherms from homogeneous and polydistributed species are delineated.

[†] ICI Australia, Newsom Street, ASCOT VALE, 3032, AUSTRALIA
[¶] Comalco Research Centre, Edgars Road, THOMASTOWN, 3074, AUSTRALIA

INTRODUCTION

The adsorption of nonionic surfactant amphiphiles at the solid-liquid interface is of undoubted importance to a large number of industrial processes. Typical examples include powder dispersion, particle stabilization, emulsification and mineral processing [1-4]. Commercial nonionic surfactants of the alkyl ethoxylate and alkyl phenol ethoxylate type are commonly used in such processes.

The mode of preparation of these commercial amphiphiles leads to a statistical distribution of ethylene oxide units in the hydrophilic moiety of a given molecular species [5]. It is of interest that the solution and interfacial behaviour of these polydisperse nonionic surfactants often approximates that of the homogeneous (pure) surfactant oligomer corresponding to the average composition of the polydisperse mixture [6].

Generally, with the exception of some anomalies shown by the first few members of the series, the critical micelle concentration (CMC), oil-water partition, cloud point, and phase inversion temperature of both the homogeneous and polydisperse polyoxyethylene nonionics are systematic, increasing functions of the ethylene oxide (EO) chain length for a given alkyl or alkyl aryl hydrophobe. Adsorption at the solid-liquid interface decreases with increasing EO chain length and parallels observations for adsorption at the air-water interface.

Despite the wealth of experimental evidence describing similarities, the behaviour of polydisperse surfactants can vary significantly from that of the corresponding homogeneous material. For example:

-   CMC's are less well defined with minima in the air-water interfacial tensions in the region of the CMC [1,7-10]. This coincides with an increase in the concentration of free monomer in solution as the total surfactant concentration (C_T) is increased above the CMC [9-12].
-   CMC's, surface tensions at the CMC, and limiting areas per molecule are lower than those for the corresponding homogeneous surfactants [6,8].

-   Oil-water partition results in oligomer distributions in the oil phase with lower average molecular weights than the total surfactant mixture with correspondingly higher average molecular weights in the aqueous phase. The distribution average in both phases decreases with a decrease in the oil/water phase volume [13-17].

-   Isotherms for the the adsorption at the solid-liquid interface are generally observed to plateau at concentrations in excess of the independently measured CMC's of the surfactants [18-28]. As the average EO chain length is increased, isotherms are observed to plateau nearer to and then below the independently measured CMC's. Maxima in adsorption isotherms have also been observed for nonyl phenol polyoxyethylenes [22,23,29,30].

Isotherms have been reported for the adsorption of polydistributed alkyl and alkyl aryl nonionic surfactants at the solid-aqueous interface for a wide variety of adsorbents. These have included silica gel [18,21,24,25,29,31-34], precipitated silica [22,23,35], alkylated silica [22,23,25,29,31,34,35], various carbonates and metal oxides, graphitized carbon black [1,35], carbon black [26,32], latex [27,28], and coal [1]. All these studies have only considered the total adsorption and have not addressed the role of the individual oligomer partition to the solid-aqueous interface. It has generally been assumed that the mixture behaves in a similar manner to the component that has the average composition of the mixture. Very few studies have controlled the surface/solution phase ratio to enable a reliable comparison of results.

The more reliable investigations of the adsorption of nonionic surfactants are those that have dealt with homogeneous single component nonionic surfactants. The adsorption of these surfactants onto hydrophobic solids such as graphitized carbon black [36-38] and alkylated silica [25] is generally characterised by a Langmuir type isotherm where the adsorbate-adsorbent interaction is, due to the shape of the isotherm, assumed to be very high. The adsorption to more polar hydrophilic interfaces such as silica gel [21,24,25,39,40] and silver iodide [41] is generally characterised by sigmoidal S-shaped isotherms which are indicative, in the limit of low surface coverages, of a weak adsorbate-adsorbent interaction. A sharp increase (inflection) is then observed for concentrations well below the CMC. This is usually associated with the onset of a strong co-operative adsorption between the surfactant molecules. Adsorption then increases rapidly to a plateau value that is usually achieved at concentrations that are less than, or equal to the independently measured surfactant CMC.

For polydistributed surfactants, Langmuir type isotherms also characterize adsorption to hydrophobic adsorbents [1,26,32,35] and sigmoidal S-shaped isotherms to more polar hydrophilic surfaces [19,20,22,23,35]. Nonetheless, there are a few essential differences to the homogeneous single component material.

- The inflection in the S-shaped isotherms associated with the onset of co-operative adsorption usually occurs at lower concentrations than for the equivalent homogeneous material and the steep portion of the isotherm is observed to be be less steep. The plateau adsorption is usually reached at 1-3 times the independently measured CMC. As the EO chain length is increased, the initial slope of the isotherm increases and the low concentration inflection becomes less prominent [23,36]. This indicates an increase in the surfactant-surface interaction [36,42]. The plateau in the isotherm also occurs at successively lower concentrations relative to the CMC.

- The area per molecule corresponding to plateau adsorption of homogeneous surfactants at the solid-aqueous interface are generally less than or equal to the corresponding limiting areas per molecule at the air-water interface [21,25,36,37,39,41,43-46]. In contrast, limiting areas per molecule corresponding to plateau adsorption for polydistributed surfactants are higher than those reported for the same species at the air-water interface.

The effect of the surface/solution ratio on the adsorption of the polydistributed surfactants has usually been almost entirely overlooked in these interpretations.

The aim of this work was to demonstrate the importance of the surface to solution ratio for polydistributed nonionic surfactants and to consider the limiting cases where a fruitful comparison might be made between polydistributed and homogeneous species. A range of surfaces were considered, including precipitated silica, low and high charge latices, methylated (alkylated) silica and graphitized carbon black. A range of alkyl aryl ethoxylates from the commercial nonyl phenol ethoxylate (ICI Teric series) range were used as adsorbates. Oligomer distributions in both the adsorbed and solution phases were determined through high resolution capillary gas chromatographic analysis.

EXPERIMENTAL SECTION

Surfactants used in this study were the polyoxyethylene nonyl phenols with a stated average polyoxyethylene chain length of 8 (Ñ8), 9 (Ñ9) and 13 (Ñ13) were supplied by ICI Operations Australia Pty. Ltd. The mean of the oligomer distribution was calculated from the results of high resolution capillary GC analysis which was found to show good linearity of results for EO chain lengths of less than or equal to nine. The distributions were found to closely resemble a Poisson distribution. These and other characteristics are shown in Table 1. The surfactants were used without further purification. CMC's were determined from surface tension measurements at 20°C.

Table 1: Characteristics of polydistributed surfactants used in this study.

Surfactant	Cloud Point (°C)	Ñ (nominal)	ÑGC	CMC (moles dm^{-3})
Ñ8	31.0±0.5	8	7.3	5.1×10^{-5}
Ñ9	50.5±0.5	9	7.7	5.8×10^{-5}
Ñ13	87	13	-	7.4×10^{-5}

The adsorbents were selected on the basis of their hydrophilic or hydrophobic character. The solids were a BDH precipitated silica, a Hopkins & Williams precipitated silica, a Spherosil XOB-015 silica gel, a methylated BDH precipitated silica, a graphitized Printex U carbon black, a low charge carboxylate latex and a high charge amphoteric latex. The BET surface area and the moisture content of each sample was measured using standard procedures. The surface area of the latex samples was calculated from sizing data assuming a density of 1.0×10^6 g m^{-3}. The surface areas and moisture contents of the samples are summarized in Table 2.

The BDH precipitated silica and Hopkins & Williams precipitated silica both dispersed readily in water and were used as received. The sample of Spherosil was manufactured by Rhône-Poulenc and was kindly supplied by Dr S Partyka, Centre de Thermodynamic et de Microcalorimetry du CNRS, Marseilles, France.

The methylated silica was prepared by reaction of 10 g dried silica with a solution of trimethyl chlorosilane in hexane. The sample was then filtered, washed with further volumes of hexane and dried at 80 °C. N_2 adsorption studies showed this sample to be quite heterogeneous by comparison with the other adsorbents.

The graphitized Printex U carbon black was prepared by heating a Degussa Printex U carbon black under N_2 at 2700 °C for 2 hours [47].

The amphoteric latex was prepared following the procedure of Homola and James [48]. Microelectrophoresis showed the sample to have an isoelectric point of 7.0 and potentiometric titration showed a density of surface groups equivalent to one charge per 0.20 nm^2. The carboxylate latex was prepared using a standard polymerisation technique [49,50]. Potentiometric titration showed a density of surface groups equivalent to one charge per 5.0 nm^2. The latices were cleaned by centrifugation followed by ultrafiltration.

All water used in this study was distilled and then passed through a Millipore 'Milli-Q' water purification system. The resultant water had a resistivity greater than 10^6 Ω/cm and a surface tension of 72.0 mN m^{-1} at 25 °C. All other reagents were AR grade.

Surfactant adsorption at the solid-aqueous interface was measured as difference between the initial surfactant concentration and that after equilibration for 48-60 hours at 20°C and pH $\approx$ 6. The solid was removed by centrifugation and the equilibrium surfactant concentration determined by UV absorption at 275 nm. The adsorptions were performed with a known surface area of solid to solution volume ratio. This ratio, R_{AV}, varied from 400 to 12.5 $(\times 10^3 \text{ m}^{-1})$.

Table 2: Characteristics of solids used for adsorption.

Solid	N_2 BET $(\text{m}^2 \text{ g}^{-1})$	Moisture $(\% \text{ w/w})$	Porosity (g cm^{-3})
BDH ppt silica	62±10	10.0	None
Hopkins & Williams ppt silica	140±16	5.0	-
Spherosil XOB-015	24±1	1.5	1[*]
Amphoteric latex (0.218 µm)	4.59	-	None
Methylated BDH ppt silica	48±4	3.0	None
Carboxylate latex (1.081 µm)	0.93	-	None
Carboxylate latex (2.25 µm)	0.45	-	None
Printex U carbon black	75±5	1.2	-

[*] Reference 21

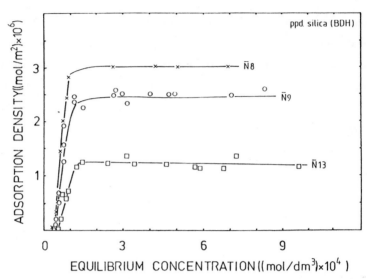

<u>Figure 1:</u>   Adsorption isotherms of Ñ8, Ñ9 and Ñ13 on a BDH precipitated silica for an R_{AV} of 200 x 10^3 m^{-1}.

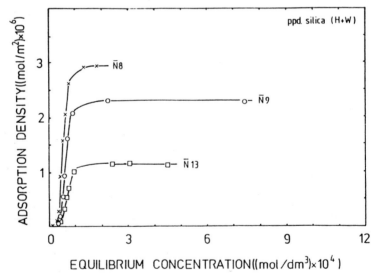

<u>Figure 2:</u>   Adsorption isotherms of Ñ8, Ñ9 and Ñ13 on a Hopkins & Williams precipitated silica for an R_{AV} of 200 x 10^3 m^{-1}.

Adsorbed nonionic surfactant was recovered by a two step procedure which involved separation of the dispersed solid from the supernatant liquor by centrifugation and followed by soxhlet extraction of the recovered solid with methanol. The liquor containing the extracted surfactant was evaporated to near dryness on a rotary evaporator and then desiccated. These samples were then dissolved in 3 cm^3 of a 50% w/w solution of hexane in acetone and prepared for quantitative high resolution GC analysis as TMS ethers [51].

Surfactant in the supernatant liquid was extracted with an equal volume of ethyl acetate. The sample was then rotary evaporated, desiccated and prepared for GC analysis. Quantification of the GC analysis was achieved by the method of internal standard additions using n-decanol (Synchemica > 95 % purity) [51].

RESULTS AND DISCUSSION

Adsorption isotherms for Ñ8, Ñ9 and Ñ13 on BDH silica, Hopkins & Williams silica, Spherosil XOB-015 silica, methylated BDH silica and graphitized Printex U carbon black are shown in Figures 1 to 5 respectively. The surface area to volume ratio (R_{AV}) is 200 x 10^3 m^{-1} in all cases. Isotherms for Ñ9 on an amphoteric latex and Ñ9 on two carboxylate latices at an R_{AV} of $\approx$ 60 x 10^3 m^{-1} are shown in Figure 6.

The isotherms are typically S-shaped for the polar adsobents, namely the precipitated silicas, the amphoteric latex and the silica gel. These isotherms are indicative of a weak adsorbate-adsorbent interaction [52]. The inflections in the isotherms for the precipitated silicas occur at concentrations approaching the CMC (typically 0.7-0.8 of the CMC) and at reduced concentrations for the silica gel (typically 0.4-0.5 of the CMC). Plateau adsorption occurs at concentrations in excess of the independently measured CMC in all cases.

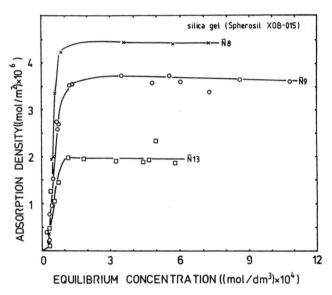

<u>Figure 3</u>:   Adsorption isotherms of Ñ8, Ñ9 and Ñ13 on Spherosil XOB-015 silica gel for an R_{AV} of 200 x 10^3 m^{-1}.

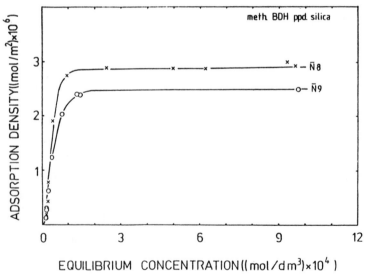

<u>Figure 4</u>:   Adsorption isotherms of Ñ8 and Ñ9 on a methylated BDH precipitated silica for an R_{AV} of 200 x 10^3 m^{-1}.

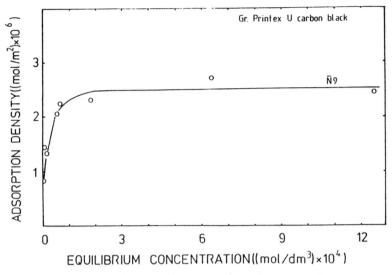

<u>Figure 5</u>:   Adsorption isotherm of $\bar{N}9$ on a graphitized Printex U carbon black for an R_{AV} of 200×10^3 m^{-1}.

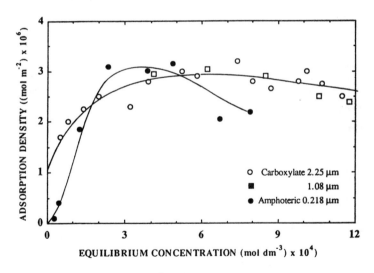

<u>Figure 6</u>:   Adsorption isotherm of $\bar{N}9$ onto an amphoteric latex and two carboxylate latices for an R_{AV} of $\approx 60 \times 10^3$ m^{-1}.

J.R. Aston, P.J. Scales, J.S. Godfrey, T.W. Healy

The isotherms for the methylated silica also have definite inflections at low concentrations. These inflections are at substantially lower concentrations relative to the CMC than for the polar adsorbents and are less pronounced. Previous results from this laboratory [23] had only shown Langmuir type isotherms for Ñ9, Ñ13, Ñ20 and Ñ40 on methylated silica. This may be due to a difference in the degree of methylation of the silicas or may be the result of insufficient data in the low concentration regime of the isotherm.

Levitz et al. [25] reported S-shaped isotherms for the adsorption of Triton X-100 (X10) and the equivalent homogeneous species on both methylated Spherosil XOB-015 and the untreated silica gel. As with the data reported here, a shift in the point of inflection to lower concentrations was observed upon methylation.

The isotherm for the graphitized carbon black and the carboxylate latex is typical of those reported for the adsorption of polyoxyethylene nonionic surfactants on hydrophobic surfaces such as carbon blacks [26,32,36-38,53-55]. There is no apparent inflection at low concentration and is indicative of strong adsorbate-adsorbent interaction [52].

The adsorption densities on these seven solids in terms of a limiting area per molecule are summarized in Table 3. The concentration at which the point of inflection is observed relative to the CMC and the plateau concentration relative to the CMC are also noted. Only the Ñ9 data is shown for clarity.

It is usually argued that at low concentrations, nonionic surfactants adsorb parallel to the surface on both hydrophobic and hydrophilic substrates [36-38,55-57]. The sharp increase in adsorption at higher equilibrium concentrations is indicative of strong co-operative adsorption between the adsorbing molecules [52] and has been associated with the formation of 'hemi-micelles' on the surface [21,24,25,34,39,56,57]. This factor and the continued adsorption of these molecules at concentrations in excess of the independently measured CMC is a consequence of the selective adsorption of the more hydrophobic surfactant oligomers from aqueous solution. This results in the solution becoming enriched in the more water soluble oligomers and results in an increase in the actual CMC of the system.

Table 3: Limiting area per molecule and isotherm characteristics for the adsorption of Ñ9 onto various solids.

Solid	$\Gamma_p \times 10^{-6}$ (mol m^{-2})	A_p (nm^2)	Inflection (x CMC)	Plateau (x CMC)	$R_{AV} \times 10^3$ (m^{-1})
BDH ppt silica	2.53	0.66	0.8	1.6	200
Hopkins & Williams ppt silica	2.28	0.73	0.8	1.7	200
Spherosil XOB-015	3.7	0.45	0.4	2.2	200
Amphoteric latex (0.218 μm)	3.1	0.54	≈ 0.8	3.9	60
Methylated BDH ppt silica	2.5	0.66	0.3	2.6	200
Carboxylate latex (1.081 μm)	2.9	0.60	0	4.3	60
Carboxylate latex (2.25 μm)	2.9	0.60	0	4.3	60
Printex U carbon black	2.5	0.66	0	2.6	200

Γ_p is the adsorption density at the plateau, A_p is the limiting area per molecule and R_{AV} is the surface area of solid to solution volume ratio used in each experiment.

In this study, the adsorption density on the Spherosil silica gel is substantially higher than on all other surfaces studied. Comparison with other work is made difficult by differences in R_{AV} between studies. Literature perusal shows qualitative agreement for the adsorption of polydisperse surfactants on similar adsorbents to those used in this study [1,19,22,23,27,28,31,35]. There are reports that show substantive differences in adsorption density. An example is the work of Abe and Kuno [26] for the adsorption of Ñ9 on a furnace black where they found adsorption densities a factor of three lower than the present study for graphitized carbon under similar R_{AV} conditions.

The only tangible reason for the difference between Spherosil and the other adsorbents appears to be a significant difference in pore volume. This could lead to condensation of the surfactant in the pore volume and significantly larger adsorption densities. Measurements of pore diameter have shown a value of 130 nm for the Spherosil sample [21]. There would be little restriction to the surfactant entering the pore volume under these conditions. Study of the other adsorbents shows them to have insignificant porosity or in the case of the latices, to be most unlikely.

Experiments were then performed to test the effect of changing R_{AV}. It was assumed this would have the same effect as changing the oil/water volume ratio in an oil/water partition experiment. Crook et al.[15] and Warr [58] have demonstrated for polyoxyethylene alkyl phenols that such a change produces an enrichment of the oil and water phases with the low molecular weight (more hydrophobic) and high molecular weight (more hydrophilic) oligomers of the distribution respectively.

Isotherms for the adsorption of Ñ8 onto the Hopkins & Williams silica at 20°C with R_{AV} ranging from 12.5 to 450 (x 10^3) m^{-1} are shown in Figure 7. All isotherms have the characteristic sigmoidal shape associated with the adsorption of these species to a polar adsorbent. However, two points are noteworthy. The limiting adsorption density increases as R_{AV} decreases and the plateau in the isotherm is achieved at higher concentrations as R_{AV} increases. Indeed, the limiting area per molecule varies from 0.37 to 0.61 nm^2 as R_{AV} increases from 12.5 to 450 (x 10^3) m^{-1} and the plateau adsorption varies from 1.11 x CMC to 1.94 x CMC over the same range. This latter trend is better demonstrated by Figure 8 where the limiting area per molecule and adsorption density are plotted as a function of R_{AV}. It is apparent on examining Figure 8 that useful comparison of isotherms for polydistributed nonionics can only be made at equivalent values of R_{AV} or at values of R_{AV} such that limiting areas per molecule approach the asymptotic limit.

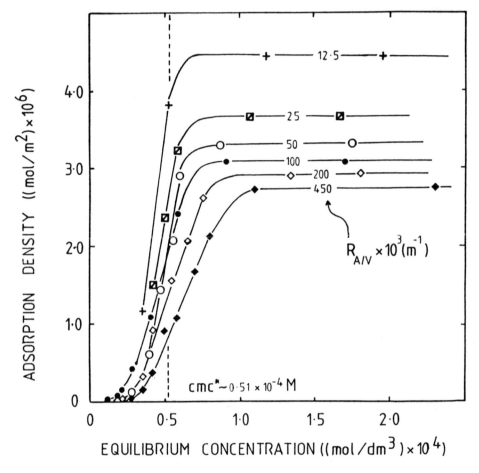

Figure 7:   Adsorption isotherms of Ñ8 on a Hopkins & Williams precipitated silica as a function of R_{AV}.

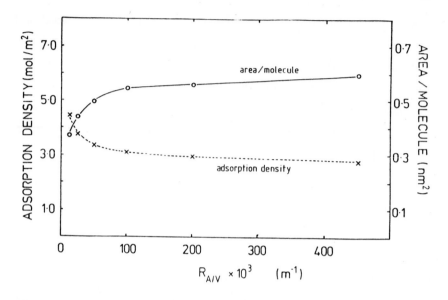

Figure 8:   Limiting adsorption density and area per molecule for the adsorption of the
             N̄8 polyoxyethylene nonyl phenol on a Hopkins & Williams precipitated
             silica as a function of R_{AV}.

Comparison of the limiting area per molecule at the solid-aqueous interface and that at
the air-water interface is also of interest. The equivalent R_{AV} at the air-water interface
will be of the order of 0.2×10^3 m^{-1}. From Figure 8, extrapolation to $R_{AV} = 0$ gives
a value for the limiting area per molecule of ≤ 0.37 nm^2 which is substantially less
than the 0.50 ± 0.3 nm^2 [58,59] at the air-water interface. Since bi- or multilayer
formation is most unlikely at the air-water interface, the conclusion must be that an
amount in excess of a monolayer is present at plateau adsorption at the solid-solution
interface. The data do not suggest however that multilayers or indeed, even a full
bilayer forms at the solid-aqueous interface. There is also no clear trend in the point of
initial inflection of the isotherms as a function of R_{AV} which suggests that there is no
significant changes in the adsorbate-adsorbent interactions.

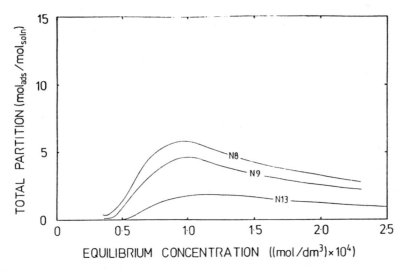

Figure 9:   Adsorption partition isotherm for Ñ8, Ñ9 and Ñ13 on BDH precipitated silica at an R_{AV} of 200×10^3 m^{-1}.

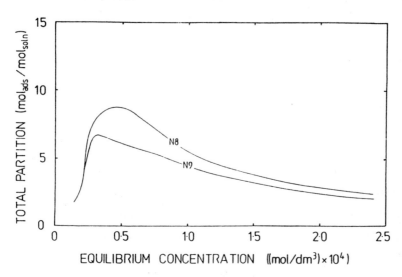

Figure 10:   Adsorption partition isotherm for Ñ8, Ñ9 and Ñ13 on a methylated BDH precipitated silica at an R_{AV} of 200×10^3 m^{-1}.

The two clear trends in the data as a function of R_{AV}, namely the change in the concentration of plateau adsorption and the limiting area per molecule at the plateau, are both indicative of adsorption producing an enrichment of the solid phase having the more hydrophobic oligomers of the distribution with increasing R_{AV}. Consequently, the solution phase will be enriched in the more hydrophilic oligomers (producing an increase in the solution CMC) with increasing R_{AV}. The affinity of the surfactant for the adsorbent as a function of chain length helps in the assessment of this phenomenon. This may be expressed as a partition function (K_{eff}) of the surfactant between the solid and solution phases. Figures 9 and 10 show the partition function for a BDH precipitated silica and a methylated BDH silica respectively. As with a homogeneous surfactant, substantial differences are observed between adsorption for different levels of hydrophobicity and for different chain lengths. The position of the maximum is however quite different from that observed for a homogeneous species.

The competing adsorption, or partition of the surfactant oligomers to the adsorbed phase was examined by monitoring the adsorption of Ñ8 to the Hopkins & Williams precipitated silica. This was achieved using high resolution capillary GC after extraction of both the adsorbed and solution phases. Measurements were performed for equilibrium concentrations below, at and well above the CMC and for an R_{AV} of 100 $x 10^3$ m^{-1}. A GC chromatogram for Ñ8 on Hopkins & Williams silica at the independently measured CMC is shown in Figure 11. Extracts from the solution and solid phase are shown. Analysis of the distributions shows an average ethoxylate chain length of 8.07 and 7.42 respectively. The differences are not immediately obvious from the chromatograms and differences in the total amount of surfactant analysed makes comparison difficult. Figure 12 shows the normalized distributions as a function of EO chain length. The aqueous phase is seen to be enriched in the more hydrophilic oligomers of the distribution and the solid phase in the more hydrophobic components. The differences nonetheless are substantially less than those observed for Ñ8 in oil/water partition studies by both Harusawa et al. [16] and Allan et al. [17,59].

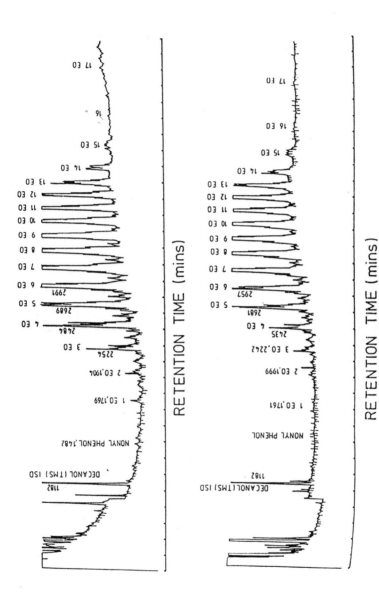

Figure 11:   Gas chromatograph of the equilibrium surfactant distributions in the adsorbed (upper trace) and solution (lower trace) phases following the adsorption of N8 on a Hopkins & Williams precipitated silica at an RAV of 100 (x 10^3) m^{-1} and an equilibrium surfactant concentration in solution of 5.2 x 10^{-5} moles dm^{-3} (1.0 x CMC).

To understand the data of Figure 12 and its relationship to the oil/water partition, it is also necessary to consider the effect of micelles in the aqueous phase upon the actual monomer distribution. Micelles are likely to be present in the solution phase for equilibrium concentrations approaching the surfactant adsorption plateau.

Table 4 shows the average EO chain length in each phase and measured partition function. It should be noted that the measured average EO chain length for the neat $\bar{N}8$ is 7.3. Also included in Table 4 is the average EO chain length in micelles and as monomer in aqueous solution. These values were calculated assuming a free energy of mixing model [9,60] and a slightly skewed Poisson distribution for the neat mixture [51]. In this model, the concentration of the individual components in the monomer and those in the micelle were calculated by substituting the experimental values for the total concentration (C_T) and for the mole fractions of oligomer in the combined micellar and aqueous phases (α_i^*) into

$$1 - C_T \sum_i (\alpha_i^* / (C_{Mi} + C_T^{mic})) = 0$$

where C_{Mi} is the CMC of a single monomer species i and C_T^{mic} is the total concentration of surfactant in micelles. The equation was then solved numerically.

The procedure provided an interesting result. At the adsorption plateau, the total concentration of monomer remains essentially constant while the total adsorption ranges from 2.95 to 4.45 x 10^{-6} moles m^{-2} for experiments with R_{AV} ranging from 200 to 12.5 x 10^3 m^{-1}. This variation is due to a variation in the concentration of the individual monomer components as a function of R_{AV}. In all cases where micelles were present, solution distribution averages were higher than micellar distribution averages in keeping with the phase separation approach to micelliszation [9,60]. It was generally noted also that the average EO chain length of the adsorbed surfactant was smaller than the surfactant present as micelles.

Table 4: Partition coefficients and average EO chain length in each phase for the adsorption of Ñ8 to a Hopkins & Williams silica at a R_{AV} of 100×10^3 m^{-1}.

Concentration (x CMC)	Γ/Γ_P	K_{eff}	Ñ (aq)	Ñ (solid)	Ñ (mon)	Ñ (mic)
0.55	0.14	1.50	7.85	7.08	7.85	-
1.00	0.78	4.15	8.07	7.24	8.13	7.27
3.84	1.00	1.82	7.64	6.98	8.37	7.46

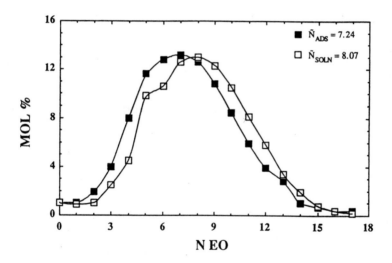

Figure 12: Equilibrium surfactant distributions in the adsorbed and solution phases following the adsorption of Ñ8 on a Hopkins & Williams precipitated silica at an R_{AV} of $100 (\times 10^3)$m^{-1} and an equilibrium surfactant concentration in solution of 5.2×10^{-5} moles dm^{-3} (1.0 x CMC).

CONCLUSIONS

The adsorption of polydisperse nonionic surfactants of the polyoxyethylene nonyl phenol type is effected by both the average chain length of the polyoxyethylene moiety and the surface to solution volume ratio. Both effect the total partition to the surface and control the selectivity of the surface for the various oligomer species. High resolution capillary GC proved useful in the analysis of the oligomer distributions in the adsorbed and solution phases and raised greatly the understanding of the adsorption phenomenon.

As a result of the possible changes to the oligomer distribution on adsorption, comparison with adsorption of the equivalent homogeneous surfactants is only suggested in the limit of the total partition approaching zero (i.e., as $R_{AV} \rightarrow 0$). Comparison of the adsorption behaviour of polydisperse surfactants is only suggested in the limit of the total partition approaching infinity (i.e., as $R_{AV} \rightarrow \infty$).

ACKNOWLEDGEMENTS

JRA acknowledges support in the form of a postgraduate research scholarship from the Commonwealth of Australia. This work was supported by the Australian Research Council. We also thank ICI Australia Pty. Ltd. for the provision of samples.

REFERENCES

1 Aston, J.R., Deacon, M.J., Furlong, D.N., Healy, T.W., and Lau, A.C.M., (1981) in Proceedings of the 1st Australian Coal Preparation Conference, (Ed A.R.Swanson), pp 358-378, Newey and Beath Printers Pty. Ltd., NSW, Australia.

2 Aston, J.R., Drummond, C.J., Scales, P.J., and Healy, T.W., (1983) in Proceedings of the 2nd Australian Coal Preparation Conference, (Ed R.L. Whitmore), pp 148-160, Westminster Press, Brisbane, Australia.

3 Rosen, M.J., (1978), Surfactants and Interfacial Phenomena, Wiley-Interscience, New York.

4 Becher, P., (1967) in Nonionic Surfactants, (Ed M.J. Shick), pp 604-625, Marcel Dekker, New York.

5 Sachat, N., and Greenwald, H.L., (1967) in 'Nonionic Surfactants,' (Ed M.J. Shick), pp 8-43, Marcel Dekker, New York.

6 Becher, P., (1967). in Nonionic Surfactants (Ed. M J Schick), p 478, Edward Arnold, London.

7 Hsiao, L., Dunning, H.N., and Lorenz, P.B., (1956), J. Phys. Chem., 60, 657

8 Crook, E.H., Fordyce, D.B., and Trebbi, G.F., (1963), J. Phys. Chem., 67, 1987

9 Warr, G.G., Grieser, F., and Healy, T.W., (1983), J. Phys. Chem., 87, 1220

10 Warr, G.G., Scales, P.J., Grieser, F., Aston, J.R., Furlong, D.N., and Healy, T.W., (1984) in Surfactants in Solution, (Ed. Mittal and Lindman), vol 2, p 1329, Plenum Publishing Corp.

11 Schott, H., (1964), J. Phys. Chem., 68, 3612

12 Warr, G.G., Grieser, F., and Healy, T.W., (1984), J. Colloid. Interface Sci., 100, 573

13 Warr, G.G., Grieser, F. and Healy, T.W., (1983), J. Phys. Chem., 87, 4520

14 Greenwald, H.L., Kice, E.B., Kenly, M. and Kelly, J., (1961). Analytical Chem., 33, 465

15   Crook, E.H., Fordyce, D.B. and Trebbi, G.F., (1965), J. Colloid. Interface Sci., 20, 191

16   Harusawa, F., Nakajima, H., and Tanaka, M., (1982), J. Soc. Cosmet. Chem., 33, 115

17   Allan, G.C., Aston, J.R., Grieser, F., and Healy, T.W., (1989). J. Colloid. Interface Sci., 128, 258

18   Van Den Boomgaard, A., (1985), PhD Thesis, Wageningen, p15

19   Kuno, H. and Abe, R., (1961), Kolloid Z. u. Z. Polymere, 177, 40

20   Ackers, R.J. and Riley, P.W., (1974), J. Colloid Interface Sci., 48, 162

21   Partyka, S., Zaini, S., Lindheimer, M., and Brun, B., (1984), Colloids and Surfaces, 12, 255

22   Scales, P.J., Grieser, F., Furlong, D.N., and Healy, T.W., (1986), Colloids and Surfaces, 21, 55

23   Furlong, D.N. and Aston, J.R., (1982), Colloids and Surfaces, 4, 121

24   Levitz, P. and Van Damme, H., (1986), J. Phys. Chem., 90, 1320

25   Levitz, P., El Miri, A., Keravas, D. and Van Damme, H., (1984), J. Colloid Interface Sci., 99, 484

26   Abe, R. and Kuno, H., (1962), Kolloid Z. u. Z. Polymere, 181, 70

27   Kronberg, B., Kall, L. and Stenius, P., (1981), J. Disp. Sci. and Tech., 2, 215

28   Kronberg, B., Stenius, P., and Ingeborn, G., (1984), J. Colloid Interface Sci., 102, 418

29   Rupprecht, H., Liebl, H., and Ullman, E., (1973), Pharmazie, 28, 760

30   Hsiao, L. and Dunning, H.N., (1955), J. Phys. Chem., 59, 362

31   Seng, H.P., and Sell, P.J., (1977), Tenside Detergents, 14, 4

32   Wolf, F. and Wurster, S., (1970), Tenside Detergents, 7, 140

33   Rupprecht, H., (1978), Prog. Colloid and Polymer Sci., 65, 29

34   Levitz, P., Van Damme, H., and Keravas, D., (1984), J. Phys. Chem., 88, 2228

35 Aston, J.R., Furlong, D.N., Grieser, F., Scales, P.J., and Warr, G.G., (1982), in Adsorption at the Gas-Solid and Liquid-Solid Interface, (Ed. J. Rouquerol and K.S.W. Sing), p97, Elsevier, Amsterdam.

36 Gellan, A. and Rochester, C.H., (1985), J. Chem. Soc. Faraday Trans. I, 81, 1503

37 Corkill, J.M., Goodman, J.F., and Tate, J.R., (1966), Trans Faraday Soc., 62, 979

38 Hey, M.J., McTaggart, J.W. and Rochester, C.H., (1984), J. Chem. Soc. Faraday Trans. I, 80, 699

39 Gellan, A. and Rochester, C.H., (1985), J. Chem. Soc. Faraday Trans. I, 81, 2235

40 Klimenko, N.A. and Koganovski, A.M., (1973), Kolloidn. Zh., 35, 772

41 Mathai, K.G. and Ottewill, R.H., (1966), Trans Faraday Soc., 62, 750

42 Klimenko, N.A., Permilovskaya, A.A. and Koganovski, A.M., (1975), Kolloidn. Zh., 37, 969

43 Rosen, M.J., Cohen, A.W., Dahanayake, M. and Hua, X.Y., (1982), J. Phys. Chem., 86, 541

44 Lange, H., (1965), Kolloid Z. u Z. Polymere, 201, 131

45 Lange, H., (1975), Tenside Detergents, 12, 27

46 Corkill, J.M., Goodman, J.F. and Ottewill, R.H., (1961), Trans Faraday Soc., 57, 1627

47 Lau, A.C.M., Furlong, D.N., Healy, T.W., and Grieser, F., (1986). Colloids and Surfaces, 18, 93.

48 Homola, A., and James, R.O., (1977), J. Colloid Intface Sci., 59, 123

49 Goodwin, J.W., (1973), Brit. Polym. J., 5, 347

50 Hearn, J., (1970), Brit. Polym. J., 2, 116

51 Aston, J.A., (1988), Ph.D Thesis, University of Melbourne

52 Giles, C.H., MacEwan, T.H., Nakhwa, S.N., and Smith, D., (1960), J. Chem. Soc., 3973

53 Klimenko, N.A., (1978). Kolloidn. Zh., 40, 1105

54   Kumagai, S. and Fukushima, S., (1976). Colloid and Interface Science, (Ed. M. Kerker), vol 4, 91. Academic Press, New York.

55   Klimenko, N.A., Permilovskaya, A.A. and Koganovski, A.M., (1974). Kolloidn. Zh., 36, 788

56   Clunie, J.S. and Ingram, B.T., (1983), in Adsorption From Solution, (Ed Parfitt G.D. and Rochester C.H.,) Academic Press, London, p 105

57   Gellan, A. and Rochester, C.H., (1985). J. Chem. Soc. Faraday Trans. I, 81, 3109

58   Warr, G.G., (1981), BSc. Hons Research Report, University of Melbourne.

59   Allan, G.C., (1984), BSc. Hons. Research Report, University of Melbourne.

60   Clint, J.H., (1975), J. Chem. Soc. Faraday Trans. I., 71, 1327

INDEX